Sustainable Urban Futures

Series Editors
Zaheer Allam, Le Hochet, Morcellement Raffray, Terre-Rouge, Mauritius
Sina Shahab, School of Geography and Planning, Cardiff University, Cardiff, UK

This series includes a broad range of Pivot length books offering accessible and applied texts designed to appeal to both practitioners and academics in the field. Pivots in the series will explore how sustainability can be achieved in Future Cities and how technology can assist in supporting sustainable transitions to better respond to the urgencies of climate change, equity needs and inclusivity aligning the two core themes of Urban Science and Future Science.

Tjark Gall · Flore Vallet ·
Laura Mariana Reyes Madrigal ·
Sebastian Hörl · Adam Abdin ·
Tarek Chouaki · Jakob Puchinger

Sustainable Urban Mobility Futures

palgrave
macmillan

Tjark Gall
Paris, France

New Delhi, India

Laura Mariana Reyes Madrigal
Gif-sur-Yvette, France

Adam Abdin
Gif-sur-Yvette, France

Jakob Puchinger
Paris, France

Flore Vallet
Paris, France

Sebastian Hörl
Gif-sur-Yvette, France

Tarek Chouaki
Gif-sur-Yvette, France

ISSN 2730-6607 ISSN 2730-6615 (electronic)
Sustainable Urban Futures
ISBN 978-3-031-45794-4 ISBN 978-3-031-45795-1 (eBook)
https://doi.org/10.1007/978-3-031-45795-1

This Palgrave Macmillan imprint is published by the registered company Springer Nature Switzerland AG
The registered company address is: Gewerbestrasse 11, 6330 Cham, Switzerland

Paper in this product is recyclable.

Preface

Why to work on mobility? Why the focus on the urban scale? Why should it be people-centred and sustainable, and what does that mean? Why urban mobility futures in plural?

Before diving into the topics, this chapter aims to outline the broader picture and motivation, situate the work in the larger discourse, and aim to provide some answers to the raised questions. Initial responses can be found in many places, including the United Nations' Sustainable Development Goals (SDGs), the Sixth Assessment Report of the International Panel on Climate Change (IPCC AR6), UN-Habitat's New Urban Agenda (NUA), the European Green Deal and Mobility Packages, as well as ample national policies and strategies.

Mobility has multiple definitions. We understand it here as the movement of people in their daily life regardless of purpose or mode of transport. This excludes mostly travel, relocation, migration, as well as social mobility, even if many touching points exist. People—like all other living beings—moved continuously and lived predominantly nomadic lives until improving farming practices and favourable conditions permitted the first more permanent settlements and cities about 10,000 years ago in today's Palestine, Lebanon, and Syria. Whilst nomadic societies continue to exist until today, most people live and operate in geographically limited spaces with mostly permanent homes and working locations. Nevertheless, mobility continues to define our life. Where we live and work depends on the proximity to other people, functions like supermarkets, economic or educational opportunity, amongst many others. Every day, the majority of over 8 billion people moves to work, school, shops, or other activities, to accompany others or just for the sake of it. This might be on foot, by bike, public transport, car, other modes, or most likely a combination thereof. Without mobility—as partially observed during the COVID-19 pandemic—normal life ceases to exist. Mobility can be seen as the arteries of humankind, crucial to keep it connected and alive. Here, the focus lies on urban mobility. This distinction is made as most people today live in cities—an ongoing trend. Further, rural mobility also has many but different challenges than those in urban areas. Therefore, this focus permits both a more targeted perspective as well as a concentration on challenges affecting most of today and tomorrow's population.

The normative foundation is the goal to transition towards *sustainable* and *people-centred* urban mobility. Following the early beginning of sustainability debates in the late 1980s, the establishment of the 17 Sustainable Development Goals (SDGs) in 2015 and their respective sub-objectives defined the sustainable development discourse since. For this work, SDG 11 *Sustainable Cities and Communities* and SDG 13 *Climate Action* are leading, whilst many of the other SDGs are directly related as well. These include, SDG 3 *Good Health and Well-Being*, SDG 8 *Decent Work and Economic Growth*, SDG 9 *Industry, Innovation, and Infrastructure*, SDG 10 *Reduced Inequalities*, and SDG 12 *Responsible Consumption and Production*, all impacted through or impacting urban mobility. Adding the lens of access to opportunities, amenities and services, nearly all remaining SDGs become directly linked to sustainable urban development and mobility. The sub-SDG 11.2 calls for 'access to safe, affordable, accessible and sustainable transport systems for all, improving road safety, notably by expanding public transport, with special attention to the needs of those in vulnerable situations, women, children, persons with disabilities and older persons' by 2030 (UN, 2015). Further, the Paris Agreement (COP21, 2015) signed in the same year as the SDGs, became a leading and binding global agreement to implement actions to address climate change in government policies. Urban mobility is strongly linked with the vows to keep 'the global average temperature to well below 2 °C above pre-industrial levels and pursuing efforts to limit the temperature increase to 1.5 °C above pre-industrial levels', by regularly measuring the levels and strategically mitigating greenhouse gas emissions whilst fostering sustainable development.

In the context of urban areas, various policies and tools followed which have been defining sustainable urban development thereafter. This includes, for example, the New Urban Agenda (NUA) which has been adopted as response to SDG 11 at the UN Conference on Housing and Sustainable Urban Development (Habitat III) in Ecuador in 2016 (UN-Habitat, 2016). The NUA calls for the 'city for all' and 'right to the city', with equal access and opportunity, in particular 'equal access for all to public goods and quality services in areas such as food security and nutrition, health, education, infrastructure, mobility and transportation, energy, air quality and livelihoods' (UN-Habitat, 2016). Further, the NUA calls for the promotion of 'age- and gender-responsive planning and investment for sustainable, safe and accessible urban mobility for all and resource-efficient transport systems for passengers and freight, effectively linking people, places, goods, services and economic opportunities' (UN-Habitat, 2016).

The last round of the IPCC assessment reports adds the more technical and scientific foundation and points towards both the contribution to Greenhouse Gas (GHG) emissions of the transport and mobility sector, as well as the potential of urban mobility to mitigate and adapt via multiple levers, noteworthily via behavioural ones (IPCC, 2022; Creutzig, 2016). These goals and objectives have been localised across the world since, for example, through national urban policies, multiple strategies and plans, including Sustainable Urban Mobility Plans (SUMPs), as well as the movement towards more mobility-targeted policies and programmes.

Jointly, these motivate and define the ambition for sustainable and people-centred urban mobility. Already hidden in some of the previous quotes, the importance is not only on the aggregated values but also on the distribution within. Less relevant for overall emissions, local exposure to mobility-related pollution, toxic materials during rare material mining, or unequal accessibility to urban opportunity of peri-urban neighbourhoods are just few examples of a highly unjust status quo. The concept of *mobility justice* focuses on this and aims for an equitable access to mobility due to its impact on economic opportunities, quality of life, healthcare, overall lifestyle, amongst others (Sheller, 2018). This dynamic is often linked to the economic status, or: 'The scarcity of economic resources not only hinders this daily mobility so necessary for poor people, it also affects the nature of their journeys' (translated from Jouffe, 2014, p.2).

Urban mobility's role for climate action, its social importance, and the challenge of its highly unequal distribution constitute together the motivation for working on sustainable and people-centred urban mobility. Lastly, the book's title uses the plural of future. This has two reasons. First, there is a high uncertainty regarding many elements of the future as we will show throughout. How many people will live where? What will be their preferences? What technologies arise? How much energy is available and at what cost? How extreme will the consequences of the climate crisis be? Each of these questions cannot be answered (yet). It is, however, expected to have significant impacts on urban mobility of tomorrow. Secondly, urban mobility is a complex challenge, also referred to as wicked problem. There is no singular solution to urban mobility. It is an ongoing negotiation and compromise between multiple objectives, limited resources, and other constraints. Hence, multiple futures are representing both the high future uncertainty leading to a multitude of possible futures as well as a range of possible preferred or aimed for future states.

Following this broad introduction, we approach these topics in six chapters. Aside from the introduction and conclusion, Chapter 2 introduces a number of future trends developments of different dimensions, followed by an overview on methods and approaches in Chapter 3. The fourth Chapter describes three case study examples that aim to respond to the trends and developments using mixed methods approaches. In Chapter 5, we reflect on the contributions and provide a list of open topics and potentials for future research. Due to the objective of the book to serve different audiences with different interests and prior knowledge, a linear structure is complemented by more flexible ways of reading. Throughout the book, stand-alone boxes introduce concepts, methods, and facts. Due to the range of disciplines with respective terms and often confusing use thereof, a glossary with key terminology supplements this work. It is used to explain theory and concepts behind relevant terms from active mobility and AVs up to low-emission zones. Whilst situated at the end, we recommend having a short look at the beginning to enable a common starting point and language.

Lastly, a few words on the origin of this work. Most of it has been conducted between 2019 and 2023 at the Anthropolis Chair, a research group hosted at the Technological Research Institute IRT SystemX and the Industrial Engineering Laboratory LGI at CentraleSupélec, University Paris-Saclay. Some of the ideas and directions build on the preceding first cycle of the Chair (2015–2019). Additional to general information and some other projects, the book contains predominantly elements of the doctoral projects of Laura Mariana Reyes Madrigal, Tarek Chouaki, and Tjark Gall, as well as of the post-doctoral research of Adam Abdin at the Future Cities Lab. The Chair will be briefly introduced at the start of Chapter 4. Aside from the purpose of contextualising the presented case studies and research foci, this further implies that for some of the work presented in Chapters 2 and 3 and each project in Chapter 4, further publications and information are available and referenced throughout. Whilst the Anthropolis Chair's second cycle will be completed by the time of publication, a number of reports, presentations, and scientific publications remain available additional to this book and await feedback, questions, and recommendations.

Gif-sur-Yvette, France

Tjark Gall
Flore Vallet
Laura Mariana Reyes Madrigal
Sebastian Hörl
Adam Abdin
Tarek Chouaki
Jakob Puchinger

Acknowledgements

The work presented here is a compilation of snapshots of seven researchers linked through the Anthropolis Chair and working towards people-centred urban mobility in one way or the other. We thank the Technological Research Institute IRT SystemX and CentraleSupélec as hosting institutions for the Anthropolis Chair, as well as the partners Groupe Renault, Nokia Bell Labs, EDF, Engie, and the inter-council partnership Paris-Saclay. Some ideas originated from the first cycle of the Chair with the additional partners of Alstom, RATP, and SNCF. Aside from collaborations within, multiple external collaborators contributed to projects referenced within this work. We thank Abdelrahman Melegy, Bernard Yannou, Hazem Fahmi, Isabelle Nicolaï, Malek Ben Ammar, Mohamed Hegazy, Patrice Aknin, Sylvie Douzou, and Yassine Benider for their support and contributions to this work. The authors express their gratitude to workshop participants and interviewees who were kind enough to share their knowledge and experience with us. Further, the presented work on Cairo has been made possible due to a research visit at the American University in Cairo, hosted by Nabil Mohareb and Sherif Goubran, and a collaboration with Transport for Cairo. Other sections resulted of the work of the Future Cities Lab. Finally, we thank the series editors Zaheer Allam and Sina Shahab, the reviewers, and the team of Palgrave Macmillan/Springer, Rachel Ballard, Naveen Dass, and Cecile Schuetze-Gaukel.

The works presented in this book are supported by the French government under the 'France 2030' programme and previously the 'Investissements d'Avenir' programme, as part of the SystemX Technological Research Institute. Some parts have been conducted at the Future Cities Lab, a joint research lab at École Centrale Pékin in collaboration with Beihang University, CentraleSupélec, and IRT SystemX with funding by the Paris Region (Region Île-de-France) and the City of Beijing.

Contents

About The Authors

Tjark Gall works on methods for people-centred and sustainable urban systems with focus on climate change impacts on urbanites and the use of urban data to strengthen evidence-based decision- and policy-making. Preceding a Ph.D. at CentraleSupélec, University Paris-Saclay, he obtained an M.Sc. Urban Management and Development Studies at IHS, Erasmus University Rotterdam, and an M.Sc. Architecture at the Technical University of Brunswick. Until 2020, he worked as project manager for the International Society of City and Regional Planners. Additional to ISOCARP, Tjark is active in the Young Academics Network of the Association of European Schools of Planning which he co-coordinated from 2021–2023 and the think tank Urban AI. Currently, he is Carlo Schmid Fellow at the Climate Change and Disaster Risk Management unit for South Asia at the World Bank in New Delhi.

Flore Vallet is leading the Anthropolis Chair since 2022 and researcher on human-centred design at IRT SystemX and Assistant Professor at CentraleSupélec. Before, she was an assistant professor at the Mechanical Systems Engineering Department of the Université de Technologie de Compiègne (UTC). She graduated in mechanical engineering design at the ENS Cachan and obtained a master's degree in industrial design from UTC. In 2012, she completed a Ph.D. on the dimensions of eco-design practices towards the education of engineering designers. She is a member of the French Eco-design of sustainable systems network and the Design Society. Her fields of interest are practices of eco-design and eco-innovation in industry and for education as well as human-centred approaches in design and eco-design. Since 2016, her main applicative field of research and academic teaching is related to design for sustainable urban mobility systems and solutions. Her work has been published in journals such as *Journal of Cleaner Production, Journal of Engineering Design, Design Studies, Design Science Journal.*

Laura Mariana Reyes Madrigal is a Ph.D. researcher on 'Mobility-as-a-Service: Concepts, governance and business models'. Mariana has a degree in Architecture and a diploma in Urban Planning and Management of Metropolitan Mobility from Mexico. She joined the Chair in November 2020, after completing an M.Sc. in Urban Planning, Transportation and Mobility at the Ecole d'Urbanisme de Paris (UGE-ENPC) with the research topic of 'Governance of Mobility-as-a-Service and its effects on public transport systems'.

Sebastian Hörl is a researcher at IRT SystemX in Paris, where he is involved in various projects on agent-based transport simulation for passenger transport and logistics. He received his M.Sc. in Complex Adaptive Systems at the Chalmers University of Technology in Gothenburg and his Ph.D. in Transport Planning at ETH Zurich. His main interests revolve around the topics of replicable use of open data and software in transport planning and applied large-scale transport simulation.

Adam Abdin is an Assistant Professor in operations research at the Laboratory of Industrial Engineering, CentraleSupélec. He is a member of the research group of Operations Management. Adam received his Ph.D. in Engineering of Complex Systems at CentraleSupélec, University of Paris-Saclay. His research is focused on developing decision support models to plan, manage, and operate complex infrastructure systems and their interactions, combining techno-economic modelling, data analytics approaches, robust optimisation techniques, and game-theory methods.

Tarek Chouaki is a Ph.D. researcher on 'Reinforcement Learning and Stochastic Optimisation for the Design of On-Demand Mobility Systems'. He joined the Anthropolis Chair in December 2019 after obtaining a master's degree in Artificial Intelligence at Sorbonne Université in Paris. Tarek focuses on agent-based mobility simulations and their usage for the design of operation algorithms for on-demand mobility systems based on autonomous vehicles.

Jakob Puchinger is professor in Supply Chain Management and Logistics at EM Normandie since 2022, affiliate professor at CentraleSupélec, Université Paris-Saclay, co-director of the Future Cities Lab with Centrale Pékin, and scientific advisor for the Anthropolis Chair at IRT SystemX. He holds a doctoral degree from TU Wien (2006), investigating the combination of metaheuristics and integer programming for solving cutting and packing problems. After his Ph.D, Jakob joined the NICTA Research Centre at the University of Melbourne working on generic hybridisation of constraint and mathematical programming techniques. He joined the Austrian Institute of Technology in 2008 where he became head of the business unit Dynamic Transportation Systems in 2014. He was Anthropolis Chair Holder and professor at IRT SystemX and CentraleSupélec from 2015 to 2022. His main research interests are in logistics and urban mobility, disruptive technologies, and the optimisation of the underlying transport systems. Jakob Puchinger is the author of more than 100 scientific publications.

Abbreviations

AV	Automated Vehicles
BEV	Battery electric vehicle
BRT	Bus Rapid Transit
CO_2	Carbon dioxide
CO_2e	Carbon dioxide equivalent
EC	European Commission
EU	European Union
EV	Electric Vehicles
FAIR	Findable, Accessible, Interoperable, Reusable
GDPR	General Data Protection Regulation
GHG	Greenhouse Gas Emissions
GPS	Global Positioning Systems
GTFS	General Transit Feed Specification
HTS	Household Travel Survey
ICE	Internal Combustion Engine
ICT	Information and Communication Technologies
IEA	International Energy Agency
IPCC	Intergovernmental Panel on Climate Change
ITDP	Institute for Transportation and Development Policy
kWh	kilowatt-hour
LEZ	Low Emission Zone
NUA	New Urban Agenda
OD	Origin-Destination
OSM	Open Street Maps
PHEV	Plug-in Hybrid Electric Vehicles
pkm	Passenger kilometre

PM	Particulate Matter
RBM	Results-based management
SAEV	Shared Automated Electric Vehicles
SDG	Sustainable Development Goal
TAZ	Traffic Analysis Zones
TFC	Transport for Cairo
TOD	Transit-Oriented Development
UE	User Equilibrium
UI	User Interface
UN	United Nations
UX	User Experience
V2G	Vehicle-to-Grid
V2I	Vehicle-to-Infrastructure
V2V	Vehicle-to-Vehicle

List of Figures

List of Tables

List of Boxes

CHAPTER 1

Introduction

Abstract The first chapter provides an overview of the motivation and key concepts used in the book. This includes the definition of urban mobility and its conceptualisation as complex system, the explanation of specific challenges, and the foresight dimension of the work towards people-centred and sustainable urban mobility futures. The chapter provides the theoretical and conceptual basis for the subsequent chapters.

Keywords Urban mobility · Urbanisation · Climate crisis · Sustainability · Inclusion · Futures · Scenarios

Building on the larger conceptual and normative framing, the introduction serves the more detailed definition of (urban) mobility and how it can be understood and modelled. For this, a systems approach is used. Further, we contrast mobility with transport(ation) and the broader mobilities field to delineate this work whilst simultaneously introducing its close relatives and elaborating on the multi-disciplinary character of urban mobility studies. Aside from mobility itself, the urban context and its implications for the mobility discussion are presented. We continue with the normative framing started in the previous section by mostly diving deeper into the environmental dimension. Finally, we expand the

T. Gall et al., *Sustainable Urban Mobility Futures*, Sustainable Urban Futures, https://doi.org/10.1007/978-3-031-45795-1_1

disciplinary rooting and develop the reasoning for working with multiple possible futures further.

This book's focus is on human, personal mobility with special attention to urban mobility. In short, *urban mobility* is used to describe *human mobility* within an *urban area. Mobility* is seen as the actual geographical change of location (mobility) or the capability thereof (motility) by humans, regardless of the mode and mean (Kaufmann, 2002, 2011). It includes the notion of the intention of moving from one point to another, as well as that of the movement for the purpose of the movement itself. This implies mobility of various categories, including individual and collective; public, private, and shared; motorised and non-motorised; commute and leisure trips, change-of-location or motion-oriented movements such as jogging. On the other hand, it excludes any movements of goods of any kind (city logistics, freight), long-term movements such as migration or nomadism, as well as social mobility (e.g., moving up socio-economic classes). Even if the latter can be considered to a certain degree as access to opportunity is defined by urban mobility and ergo directly impacts social mobility.

Here, *urban* refers to the mobility in an urban area, defined as an area with a certain population density (varying by context) and the existence of societal, cultural, economic, and administrative functions. The context of Paris, like most other larger cities, is that of an urban or rather metropolitan mobility area. This refers to a connected and continuous urban area which jointly constitutes a functional metropolitan area, determined foremost by the daily territories of its permanent and temporary residents (for Functional Urban Areas see OECD, 2019, cf. 'Bassins de mobilité' in French 'Code de Transport', Article L1215-1). Aside from the challenge to define these clearly, it must be anticipated that the area will change over time and has different extents today compared to those of studied future situations.

Combining the two concepts, we define *Urban Mobility* as the actual or intended human movement within or from and to an urban area, regardless of it being local, national, or international. As introduced before, mobility and urbanisation, as well as sustainable urban development, are intrinsically intertwined. Cities arose as marketplaces and nodes in an interconnected network, with the movements in between them and within defining their development and growth (cf. agglomeration economies in Glaeser, 2010). The increasing pace and reach of modes of mobility allowed for the spatial expansion of cities, as well as the globalisation of their markets and residents' network. Spaces and streets as physical manifestations of mobility defined the layouts and spatial patterns of cities. Until today, highways, railways, or major roads are cities' arteries and barriers. Various concepts are combining urban development and mobility directly. Mobility is one of the determining factors for the access to housing, amenities, services, or transport itself. Throughout this book, we refer to it primarily as *access to opportunity* or *access to the city*, understanding a city as container of opportunity. Within this space, various types of mobility exist. A number that significantly expanded in diversity over the recent years, including shared, collective, active, and autonomous mobility. To make sense of this, complex systems thinking permits to model urban mobility as complex socio-technical systems as shown in Box 1.1 to represent and systematically work with this diversity of the system components.

Box 1.1 Urban Mobility System model [concept]

This box introduces a three-layered urban mobility system model. Below, a visual urban mobility model representation is shown, resulting from literature, expert interviews, and workshops conducted within the scope of the Anthropolis Chair. All components can be categorised in three layers of people, infrastructures, and mobility services. This tripartite organisation helps to adapt such systems. First, the people layer can be adapted by integrating, for example, population growth. The increased population requires changes of the built environment and infrastructures, such as building new roads or public transport rails. The combination of the first two layers enables and requires various mobility services, such as bus schedules or bike sharing services.

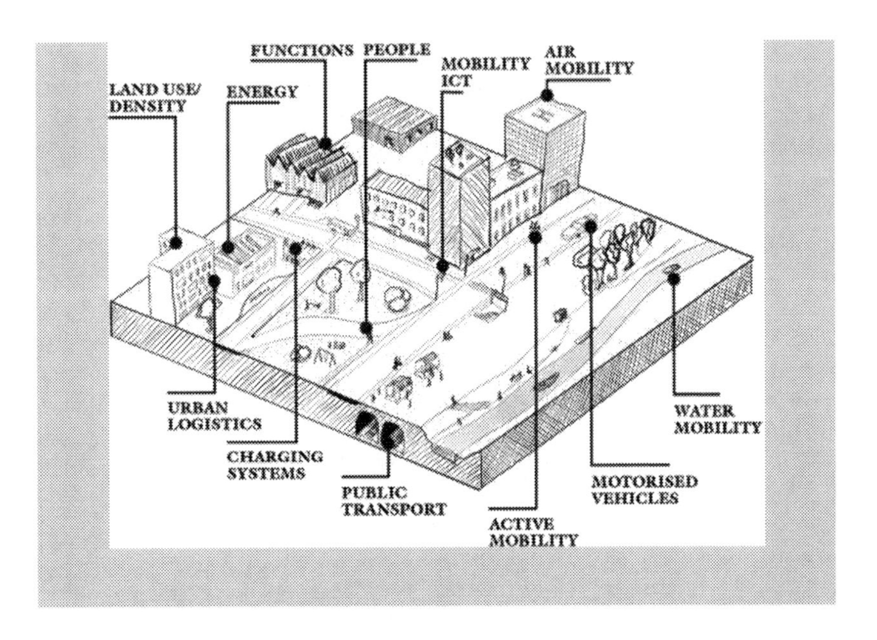

Aside from mobility, other terms are commonly used. Mostly, transport, transportation, and mobilities are all related yet distinct concepts. Transport refers to the physical act of moving something from one place to another. Initially, goods were transported and dominated any discourse. This field was extended to human mobility with more or less success. A field from which inspiration was drawn is that of pipes for water as conceptual model for traffic flow and name giver for many of its dynamics (see Dyson and Sutherland, 2021). These analogies might have helped initially but bear various risks. The most common example is that of *induced demand* which refers to rising traffic resulting from increased infrastructure capacity. Many times, when too much traffic congests the roads, more lanes are added, or new highways are constructed. However, after a short period of improved service, the demand matches the new offer, and the same congestion follows. Many of these challenges origin of the theoretical foundations from fields where the 'good' transported does not react. However, humans are not goods but instead behave and react to their environments. The increasing use of mobility instead of transport attempts to make this distinction. Transportation, on the other hand, refers rather to the process of transport. In practice, these two terms are often used interchangeably. Within this context, we talk about mobility when referring to humans moving, public transport as the service of transportation enabled by public authorities, and logistics for referring to the transport of goods. Mobilities, on the other hand, can be understood as a larger multi-disciplinary concept that encompasses the movements of people, objects, information, ideas, and cultures across various spaces and contexts. It goes beyond the traditional focus on physical movement and considers the broader dimensions. Mobilities studies explore how different forms of movement are interconnected and how they shape societies, individuals, and environments (Urry, 2002, 2007; Sheller and Urry, 2006, 2013). The emphasis is that mobility is not only about getting from one place to another but also involves complex interactions, networks, and impacts on individuals and their lives. Whilst supporting the transition of the notion 'transport' to 'mobility' as well as the research on mobilities, we call for attention to not confound mobility and mobilities or try to replace the former by the latter as arising as trend. Mobility refers to an action or potential thereof and remains primarily a physical activity which is at the core of emissions, social interaction, and many other objective realities. Mobilities studies can and should complement and nourish research on mobility but cannot replace it. In simpler terms, here, transport is the broader term to define moving *things*, mobility is the process of *people moving*, the earlier mentioned concept of motility the *capacity* thereof, and mobilities a zoom-in on *social dimensions and dynamics* linked to various types of mobility.

After this short excursion into terminology, we return to the relevance of improving urban mobility. The most tangible one is the environmental dimension of sustainability on which we focus here. GHG contributions of the transport sector are with 22% at the second place after heating, whilst 1.35 million people are killed on roadways annually (Climate Watch, 2020; CDC, 2020). Research suggests that at least 8% of global emissions result from urban mobility alone (Creutzig, 2016). Compared to other key emitting zones, real emissions are so far not decreasing significantly. For example, Germany's green transition is severely slowed due to the transport sector, notably people's mobility. Aside from emissions, road infrastructure takes up significant and valuable portions of land, i.e., 27% of space in Paris of which 57% is only dedicated to cars (Héran and Ravalet, 2008). The production of vehicles for mobility further requires a variety of scarce resources (Metabolic, 2019), and creates negative externalities for the society through lost productivity due to traffic, estimated at up to 11 billion euros per annum in Paris alone (CEBR, 2014). The impacts of over a century of fuel-based vehicle transportation are globally significant for societies today (e.g., through air pollution and sea level rise) and are expected to be even more so in the future.

This, amongst other negative externalities, combined with the crucial role urban mobility plays, requires a sustainable transition of urban mobility systems. How this can be achieved is a question we attempt to respond to from different angles throughout this book without claiming to prove a holistic response. Three assumptions are underlying and important to point out. First, mobility appears to be crucial for humankind and not desirable to be eradicated (without excluding possible advantages of working-from-home or similar practices). Second, mobility requires energy. Or in the words of Newton's second law, 'an object will not change its motion unless a force acts on it'. Building on that, the third assumption states there will be no unlimited or 100% clean energy in the near future that would permit unlimited movement. The combination of these three assumptions constitutes the core of the environmental challenge of urban mobility but is also part of the solution. A crucial determinant is the energy needed for a movement. This is constituted by distance, frequency, weight, and speed, and thus also the mode of transport. We zoom in further on distance and frequency later. Here, the focus shall be on weight, speed, and mode as highly connected concepts. First, the more weight is transported, the more energy is used. The faster this takes place, the more energy demanding. These combinations are strongly linked to the mode. Consequently, much importance is given to the mode of transport and shifting towards less polluting and energy-demanding ones. Box 1.2 showcases a simplified model what modes fall in which categories.

Box 1.2 Mobility pyramid [concept]

Adapted from Mobilitätspyramide in 2010 by German initiative 'Netzwerk Slowmotion' by Martin Held und Jörg Schindler and visualised by Ingrid Torn, Tutzing, the mobility pyramid shows that active mobility should provide the basis for most trips and only if and when needed, we can move up towards other modes. This does not imply an exclusion of one mode or the other but rather a normative prioritisation.

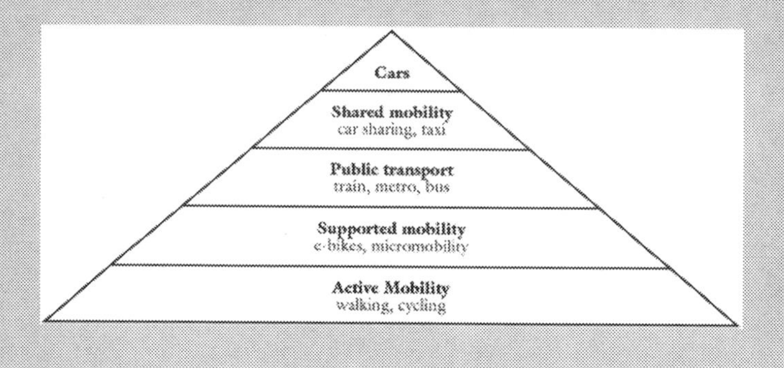

Lastly, we briefly introduced before the notion of futures and justified its plural use. We can use the visual representation of the futures cone as a representation to explain some of the key concepts (Box 1.3). We are at the present point in time where there is a certain understanding of the status quo. However, the farther we move towards the future, the more variations are possible. The cone's size increases over time. Within, non-linear pathways show possible developments, impacted by various events. At any future point in time, we can look at a section of the cone. Theoretically, an infinite number of future states exists, mostly when imagining the cone as being defined by more than two dimensions as it is the case in the visualisation. Therefore, scenarios that can be used to categorise future states into probable, plausible, possible, and preferred scenarios. They act as intermediary design objects that can reduce the complexity of futures to a manageable degree. We use the definition by Spaniol and Rowland (2018), stating that scenarios must be multiple, possible, and distinct alternatives which are rooted in the future and include some kind of narrative description.

Box 1.3 Futures cone [concept]

The futures cone is the leading visual representation of the increasing scope of future possibilities over time (Gall et al., 2022). From a singular moment today, various pathways lead to different probable, plausible, possible, preferred, or even preposterous futures. Scenarios are a set of future states at any point in time, represented by the small grey circles.

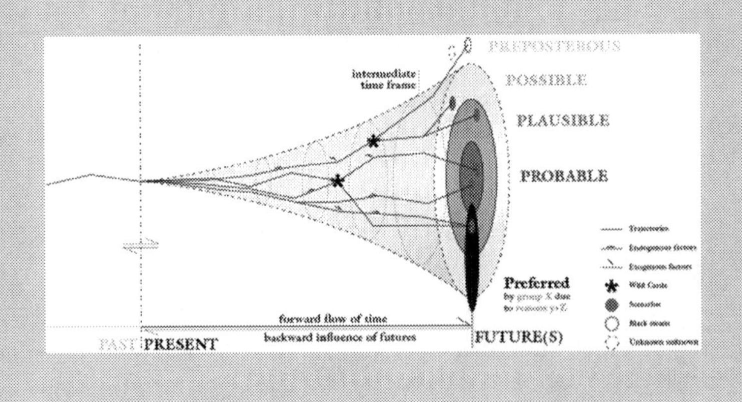

Summarising, this chapter introduced a more detailed definition of urban mobility, related concepts, and its stakes, as well as scenarios as concept to integrate futures and foresight. The next chapter looks at the developments that might happen in the future, organised in three categories of trends: Societal, urban, and technological ones.

References

CDC. (2020) Road traffic injuries and deaths—A global problem. Accessible at: https://www.cdc.gov [accessed August 2023].

CEBR. (2014) The future economic and environmental costs of gridlock in 2030. An assessment of the direct and indirect economic and environmental costs of idling in road traffic congestion to households in the UK, France, Germany and the USA. Centre for Economics and Business Research (CEBR).

Climate Watch. (2020) Historical GHG emissions. Accessible at: https://www.climatewatchdata.org [accessed August 2023].

Creutzig, F. (2016) Evolving narratives of low-carbon futures in transportation. *Transport Reviews,* Vol. 36/3, pp. 341–360. https://doi.org/10.1080/01441647.2015.1079277.

Dyson, P., and Sutherland, R. (2021) *Transport for humans. Are we nearly there yet?* London Publishing Partnership.

Gall, T., Vallet, F., and Yannou, B. (2022) How to visualise futures studies concepts: Revision of the futures cone. *Futures*, 143, p. 103024. https://doi.org/10.1016/j.futures.2022.103024.

Glaeser, E. (ed.) (2010). *Agglomeration economies.* Chicago: The University of Chicago Press/National Bureau of Economic Research.

Héran, F., and Ravalet, E. (2008) La consummation d'espace-temps des divers modes de déplacement en milieu urbain : Application au cas de l'Ile de France. Lettre de commande 06 MT E012. Ministère des transports, de l'équipement, du tourisme et de la mer.

Kaufmann, V. (2002) *Re-thinking mobility. Contemporary sociology.* Hampshire: Ashgate.

Kaufmann, V. (2011) *Les Paradoxes de la Mobilité. Bouger, s'enraciner*, 2nd ed. Lausanne: Collection Le Savoir Suisse.

Metabolic. (2019) *Metal demand for electric vehicle: Recommendations for fair, resilient, and circular transport systems.* Amsterdam: Metabolic.

OECD. (2019) Functional urban areas France, Regional Statistics, January 2019.

Sheller, M., and Urry, J. (2006) The new mobilities paradigm. *Environment and Planning*, Vol. 38/2, pp. 207–226. https://doi.org/10.1068/a37268.

Sheller, M. (2013) Sociology After the Mobilities Turn from: The Routledge Handbook of Mobilities Routledge.

Spaniol, M. J. and Rowland, N. J. (2018) Defining scenario. Futures & Foresight Science, Vol 1/1. https://doi.org/10.1002/ffo2.3

Urry, J. (2002) Mobility and proximity. *Sociology*, Vol. 36/2, pp. 255–274. https://doi.org/10.1177/0038038502036002002.

Urry, J. (2007) *Mobilities.* Hoboken: Wiley.

CHAPTER 2

Future Trends and Developments for Urban Mobility

Abstract The second chapter describes what is meant by trends, how they can be analysed, and how they relate to sustainability transitions. Further, it looks at ways at how we can change, maintain, adapt to, or strengthen them. On a content level, the focus lies on three types of trends: Societal, urban, and technological trends. The trends are taken up in the case studies of Chapter 4.

Keywords Future trends · Uncertainties · Society · Urban development · Mobility innovation

This part—the first of four main chapters—describes what is meant by trends, how they can be analysed, and how they relate to sustainability transitions. Further, it looks at ways of how we can change, maintain, adapt to, or strengthen them. On a content level, the focus lies on three types of trends: Societal trends, urban trends, and technological trends.

Designing mobility solutions or services today requires having an understanding of what is to come. Even smaller mobility interventions such as bike sharing services are usually supposed to serve the population for at least five to ten years. Other interventions, such as a new road or metro line, is planned to be in use for at least a couple of decades and can oftentimes be seen in the *urban fabric* for centuries. To design

T. Gall et al., *Sustainable Urban Mobility Futures*, Sustainable Urban Futures, https://doi.org/10.1007/978-3-031-45795-1_2

adequate services and make adequate decisions today, we must have an understanding of what the future conditions might be. When formalising them, we can distinguish between *trends* and *uncertainties*. Trends are future developments that we are relatively certain to become true. For example, population growth and urbanisation in most of the global south is relatively certain and can be understood as a trend. On the other hand, the evolution of many factors is impossible to know for certain today. For example, what will be the cost of energy in ten years, how many cars per household will there be, or how far developed is hydrogen energy or level five autonomy of vehicles? These can be referred to as uncertainties. Many uncertainties result from human behaviour (Bevan 2022) whilst others are only indirect consequences, e.g., from climate crises or changes of nature. We are interested in uncertainties with a systemic, highly intertwined, and chaotic character as they are the most uncertain and impactful. These are also referred to as *critical uncertainties*.

Trends and uncertainties together define the conditions in what ways urban mobility systems evolve and develop. We call this *system transitions*. There are many different theories around this, but here, only a few core elements are introduced. Generally, a system transition is the process of one system state to another. For example, the transitions of Paris' urban mobility system from 2015 to 2020. If we actively want to impact this process, we speak about a *system transformation*. In this process, additional to trends and uncertainties, we also need to consider the behaviours and characteristics of the urban mobility system itself. An example for a behaviour is that a congested road may lead to less demand, alternative sub-optimal route choices, or a shift to alternative *mobility modes*. An important system characteristic is the *resilience* of the overall system or its components. This refers to the capacity to withstand or absorb shocks or changes. The simplest reference is the ability to withstand disasters. For example, is a road resilient against possible landslides or floods? In more complex situations, resilience can refer to the way behaviours return to previous states, for example, after an oil crisis or the COVID-19 pandemic.

To summarise: When we want to design elements of the urban mobility system today, meaning *transform* the current system, we need to know what key *trends* and *uncertainties* there are, what the *characteristics* and *behaviour* of the system are, such as its resilience, and what *transition* are possible. A remaining question is how we want to transform the system. Meaning, what is the outcome we want to get to. It is impossible to define a specific system state today we could undisputedly call optimal for everyone. Rittel and Webber (1973) defined this as *wicked problem* without clear solution, no stable system state, no clear end point, and complex interactions. Despite the challenge to define a clear and singular preferred system state for wicked problems, we can use some of the concepts introduced before to guide our choices, as well as a commonly applied framework to reduce negative environmental impacts.

Building on the already introduced notion of sustainable and people-centred urban mobility, we cannot clearly define which solutions or pathways are the correct ones. However, referring to the SDGs and leading normative frameworks, preferred future states of urban mobility should (1) minimise emissions, (2) maximise accessibility for opportunity, and (3) ensure basic mobility capabilities for everyone, e.g., a minimum level of service level of public transport (cf. Nussbaum 2003; Sen 1979). A helpful framework to organise and prioritise mobility solutions is the Avoid-Shift-Improve (ASI) framework developed originally in Germany with uptake by GIZ, Slocat, and TUMI (Bongardt et al. 2019). The central idea is to foremost *avoid* trips that are not necessary, e.g., by digitising administrative procedures or more localised urban and land use planning where most needs can be met in the immediate environment. The remaining trips shall *shift* modes from less to more sustainable ones. For example, using the bus instead of the car or the bike instead of the metro (cf. energy demand of different modes depending on weight and speed). Finally, for trips that cannot be avoided or shifted, the transport modes shall be *improved*. For example, by improving the efficiency of engines or shifting to electric vehicles. Box 2.1 shows some examples of solutions and further information about the ASI framework. Whilst this framework does not address questions regarding accessibility, it supports a prioritisation of environmentally sustainable solutions and highlights the potential of urban planning and behavioural measures compared to technological innovation. Keeping in mind the transition context, the relevance of future trends and uncertainties, and the A-S-I framework as one way to structure existing and planned urban mobility solutions, the next sections focus on future trends and developments.

Box 2.1 Avoid-Shift-Improve Framework

The A-S-I framework enables prioritising avoid actions which reduce the need for mobility, followed by modal shifts from less to more sustainable modes, before improving what is left. The key message behind is that large gains result from behavioural changes rather than technological fixes (Bongardt et al. 2019; cf. Creutzig 2016).

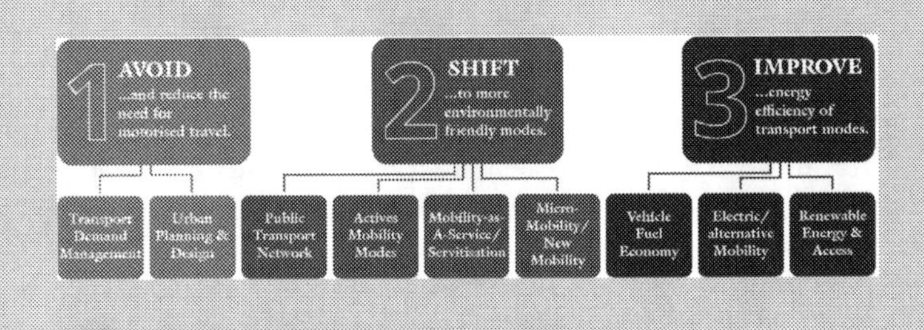

2.1 Societal Trends

In the first section of this chapter, we introduce societal trends that can have an influence on the urban mobility system. We talk about them as trends for simplicity purposes, even if their specific manifestation can vary and thus also been sometimes understood as an uncertainty. Further, societal trends are very context-depending and vary across locations. However, many similarities can be found across regions, even if often with a temporal misalignment. For example, fast urbanisation has set on first in Europe and Northern America, followed by East Asia, South America, and the MENA region, whilst now strongest in South-East Asia and Africa (UN-Habitat 2022).

Population Development and Ageing

The most important trend is demographic development. The global population is anticipated to peak at around 2100, however, with very different trajectories and stages across places. The overall growth can be best described by the number of children per person. To keep a stable population, roughly two children per woman would be needed ignoring changes of life expectancy and premature or unnatural causes of death. The importance for urban mobility system is that the population number in general defines the urbanisation and thus where people find a place to live and work, but also because different groups have different need profiles. For example, a place with a large population of the elderly (such as many Western European cities) has a different mobility demand than a place with a dominant youth population.

Between 2015 and 2030, the global number of senior citizens is expected to increase by 56%, from 901 million to over 1.4 billion (Eendebak and WHO 2015). There are countless challenges posed by the demographic shift and the market development potential of the so-called 'silver economy' seems immense. These issues of developing adapted products and services and controlling the expenses generated by an ageing population are currently at the forefront of the European discourse. It is important to not take a persisting image which associates ageing solely with a decline (Dankl 2017). Increasing quality of healthcare, life expectancies, and active senior citizens lead foremost to different use and need profiles in the mobility context, and only secondly to the challenge of moving people with reduced abilities due to age. A higher share of elderly might result in more leisure trips, higher prioritisation of comfort over speed, and different financial capacities.

Behaviours and Practices

Another component is what people do, both on an individual level and societal level. When we talk about individual behaviour, we are referring to the actions undertaken by individuals, whilst shared behaviours are categorised as practices, or social practices (see Hargreaves et al. 2012, for comparison). Behaviours encompass the actions carried out by individuals, whilst practices involve shared activities conducted by a group within society. One example is the growing trend of individualisation, denoting the increasing focus on individual preferences and choices, particularly evident in most Western societies. This trend is evident in decisions related to housing, such as suburbanisation, and the concept of individual car ownership, which symbolises a sense of freedom and autonomy. This shift in perspective is coupled with an upsurge in flexibility, as individuals detach themselves from fixed locations, participating in global movements, and adopting a more nomadic lifestyle.

Another trend is that of *tele-everything* referring to working-from-home or generally outside the office and getting food, groceries, and other products and services delivered. This trend started at different times in different locations over the past decade and accelerated everywhere strongly due to the COVID-19 pandemic. Whilst digitalisation was a barrier in some areas, e.g., having sufficient network speeds or access to smartphones, this barrier is being eradicated rapidly and globally. On the other hand, a slight return towards pre-pandemic levels can be observed whilst any equilibrium appears uncertain so far. The key impacts on urban mobility are threefold: (1) Less or no commute impact both home and office location choice. (2) Less work commutes might reduce demand, shift it to other times, or to other locations. (3) Increasing exodus of businesses from central locations and subsequent vacant spaces impact the property market and might shift urban centres or lead to increasing residential and support functions in previously business-dominated areas. It is too early to make clear statements or estimations about the overall long-term impact or the specific one of COVID-19. However, it is safe to say that it is a trend that significantly impacts urban mobility and urban dynamics in general already today and even more so in the future. Nevertheless, the possibility to work remotely still only applies for a smaller part of the overall workforce and should not be overestimated either.

Further, there is the concept of *servitisation*, representing a decrease in the emphasis on ownership. This trend exhibits significant geographical diversity, manifesting differently across various regions. In certain Western societies, the notion of 'everything as a service' gains prominence. This notion will be explored further in relation to mobility services. This transformation not only introduces distinct business models across sectors but also triggers a re-evaluation of fundamental urban services, their accessibility to the public, and the evolving role of the private sector in tandem with collaborative endeavours. In mobility, the most dominant version thereof is that of Mobility as a Service (MaaS) and itinerary calculators, raising pertinent questions about information as a potential barrier to achieving sustainability objectives.

The interplay between individual behaviours and social practices has a profound impact on the design and functionality of urban mobility systems. Studying, measuring, and trying to understand these dynamics provides a comprehensive foundation for addressing challenges, harnessing opportunities, and steering mobility solutions towards greater sustainability and efficiency.

Gender

The evolving landscape of gender roles has emerged as a significant factor shaping the dynamics of urban mobility. Over the past century, transformative events such as world wars and the emancipation movement led to substantial changes, including a rise in the number of women participating in the workforce outside homes. However, this increased participation has not led to a complete redistribution of domestic responsibilities, with women still predominantly occupying roles related to household and care work. This duality has contributed to the complexity of travel patterns, characterised by trip chains that encompass both professional and caregiving duties. For example, dropping children on the way to work, getting groceries after work, etc. Another crucial dimension arises from the heightened vulnerability of women, mostly in public transportation and during night-time travel. This calls for more diversified perspectives, transcending the conventional notion of a uniform passenger. Embracing the heterogeneous needs resulting from diverse gender roles is crucial in redefining urban mobility strategies to be more inclusive, responsive, and aligned with the diverse realities of modern society (Law 1999; Hanson 2010).

Hypermobility

The concept of *hypermobility* describes the result of increasing speed, distances, access, and frequency of mobility which followed the constantly evolving means of transport capabilities. This development is sometimes termed as out of scale or pace for humans, leading to a disconnection and growing disparity. Further, Khisty and Zeitler (2001) point out an imbalance which came along with this development, in particular, that the expansion of motorised transport networks for higher speeds resulted in traffic jams which have significantly reduced mobility, accessibility, and business productivity, increased fuel consumption and pollution and led to a widespread loss of productive time. Whilst studying its implications on time, space, human freedom, and social justice under an ethical, systemic perspective, the authors conclude 'that if hypermobility is not dealt with both as an individual and as a collective responsibility, the challenge to transport ethics and its systemic issues could be further impaired' (Khisty and Zeitler 2001).

A second related notion is that of the *hypermobile* as a group which travels disproportionally more than other groups, both locally and internationally across means of transport and consequentially has a significantly higher contribution to the environmental impact, as well as the overall mobility demand (Gössling et al. 2009). A distinction can be made between the production-oriented and consumption-oriented hypermobiles. The first describes those primarily travelling for work-related reasons, whilst the second travels mostly for consumption and leisure. Gössling et al. (2009) studied this phenomenon further based on a travel behaviour study from France. Some of their main observations of the *hypermobile* cluster were that 'citizens with higher income levels (over €7,500 per month) are overrepresented in the cluster', holding 'management positions and [being] workers with higher education' in the age group 50–69. Further, singles as well as childless couples, living in Paris, constitute a majority of the group. In numbers of impact, the study showed that 50% of the total mobility-related GHG emissions were produced by the 5% hypermobile (Gössling et al. 2009).

Whilst much of this contribution originates from international air traffic and is hence less relevant for the discussion of urban mobility, two conclusions can be drawn. First, a small group has a significant systemic impact. Hence, an in-depth understanding of this dynamic and an accordingly adapted consideration is crucial when assessing environmental impacts of changing systems. Further, the proportional distribution of different mobility behaviours and their development over time will have a significant impact. If globalisation and increasing socio-economic status would lead to a significant increase of the hypermobile population, its impacts of urban mobility futures must be taken into consideration, as well as anticipated and planned for.

Health: Sedentary But Rushed Lifestyles

Unhealthy diets, physical inactivity, and stress are central *behavioural risk factors* of cardiovascular diseases according to the World Health Organization. On one side, the lack of physical activity related to a certain type of urban lifestyles is usually linked to work routines including long hours sitting in vehicles, offices, or the home office. The lack of physical activity seems to be the direct cause of at least 7.2% of all-cause deaths and 7.6% of cardiovascular diseases-related deaths worldwide (Katzmarzyk et al. 2022). On the other side, the food choices available to consumption in cities have integrated more hypercaloric and hyper-processed meals. Healthy diets including a balanced amount of non-processed food also represent an important expense in relation to the income of certain socio-demographic groups. In car-based societies, vehicles encouraged the increasing of distances travelled (for example, by getting more affordable housing in the peripheral or rural areas of an urban core where jobs are), whilst maintaining a relatively stable travel-time budget (Marchetti 1994; Dong et al. 2022).

Just in the Île-de-France region around Paris, for every 100,000 inhabitants, 254 are estimated to die from diseases related to the circulatory system (Insee 2022). If these numbers are monetarised and internalised, the interest to fight them becomes clear for policymakers and public authorities. Specially given that there are numerous studies showing the benefits of increasing the distances walked or cycled to reduce the prevalence of heart, cerebrovascular diseases, and other mental illness. For example, Woodcock et al. (2009), identified that 'increase[ing] in the distances walked and cycled would [...] lead to health benefits [...] from reductions in the prevalence of ischemic heart disease, cerebrovascular disease, depression, dementia, and diabetes' (Woodcock et al. 2009, p. 1930).

As a way to take action in France, the Public Health Code[1] gave general practitioners the competency to prescribe adapted physical activity to their patients. This prescription is given for a duration of six months where individuals are offered the possibility to transform their health habits by doing more physical activity guided by professionals and under medical supervision. General practitioners suggest and prescribe the 'type of activity, duration, frequency, and intensity that are specified on a specific form as defined by the decree of the Ministry of Health'.

It was in 2016 that this policy was launched with the '*Law of modernisation of the French health system*'. The policy was revised and updated in

[1] FR: Code de la Santé Publique, Article L1172-1.

2022 in the '*Law for the democratisation of sports of 2022*'.[2] The legal framework aims to create awareness about the negative effects of sedentariness and at the same time, balance the degradation of the health of patients living with overweight, obesity, arterial hypertension, malnutrition, physical inactivity, dyslipidaemia, and addictive behaviour, amongst other chronic diseases listed in the Decree D1172-1-1.

This section introduced a range of societal trends, each with multiple impacts on the functioning of urban mobility systems and the requirements for solutions for the future.

2.2 Urban Trends

Aside from the societal dimension, we focus on mobility within urban areas. An introduction of a couple of concepts and trends in the urban realm shall contextualise both the status quo of urban mobility systems, as well as the anticipated ones for the future. We focus on some specific concepts such as Transit-Oriented Development, but also general trends like urbanisation, sprawl, secondary city/hinterland growth. A distinction must be made between different types of urban areas which are mostly linked to geographical location. Large cities continue to grow whilst some small cities in the western world struggle increasingly with decreasing population. Latin America as the most urbanised continent has found somewhat of an equilibrium whilst both small and large cities continue to grow in Asia and Africa, resulting in the highest future mobility demand and connected challenges. To organise types of urban settlements, the IPCC suggests categories of urban settlements. Building on the established relationships of urban form to emissions, these categories shall help to prioritise mitigation actions depending on the context. The summary report for urban policymakers states that 'settlements that are in early stages of urbanisation with relatively low levels of infrastructure deployment have large opportunities to pursue low- or net-zero urbanisation pathways, whereas established cities with mature infrastructure generally have more locked-in energy behaviours' (Babiker et al. 2022). Three typologies are proposed: an established city, a rapidly growing city, and

[2] *Law 2022–296 from 2 March 2022 to 'democratise sports in France'.*

an emerging city. The two case studies in this book, Paris and Cairo, fall into two of the categories. Paris is mostly an established city whilst Cairo remains to be rapidly growing. However, it is possible to distinguish further between different areas within a city. The categorisation is not meant as fixed concept but as mental support model to distinguish urban areas, their challenges, and potential solutions by shared characteristics. Most of the following trends apply differently across these categories.

Urbanisation

The first global trend of ongoing urbanisation led to urban mobility being a topic of interest in the first place. Whilst the world urbanised for centuries, combined with the overall population growth and peaking urbanisation growth rates in the past century, urban areas are today home to 55% of the global population. As shown in Box 2.2, by 2050, an additional 2.4 billion urban dwellers are expected. At the same time, more megacities of over 10 million inhabitants arise, most rapidly growing in the global south and creating various challenges, ranging from inequality to self-planned settlements with lacking services and amenities. Already today, urban areas are, in many cases, more powerful compared to their respective countries, making urban areas the power epicentres of today and most likely even more so in the future.

Box 2.2 An Urbanising World [Facts]

The world continues to urbanise, as detailed in the highly recommended and freely accessible publication 'Cities change the World' by L'Institut Paris Region (2019). Whilst in 2000, only 2.9 billion people lived in cities, by 2050, about 6.7 billion urbanites are expected. Also, the size of cities increases. In 1975, 4 cities had over 10 million inhabitants. By 2030, it shall be 43. These urban areas are at the same time a challenge and a solution space. They only take 2% of the land surface to house currently about 55% of the global population but emit 70% of global CO_2 emissions and consume 78% of energy (United Nations, World Urbanization Prospects: The 2018 Revision; UN Habitat, Working for a better urban future: annual progress report, 2018; LIU, Z., HE, C, ZHOU, Y. et WU,

J. in Landscape Ecology, 2014; United Nations, https://www.un.org/en/climatechange/cities-pollution.shtml, 2019; UN Habitat, World Cities Report, 2016; in: L'Institut Paris Region 2019).

Three considerations are important to be pointed out. First, challenges, innovations, and developments of urban mobility in metropolitan and megapolitan regions will define the lives of the majority of the global population, at least in the near future. Second, many urban cores are already highly densified with developable land becoming a scarce resource. This leads to urbanisation being accompanied in many cases by urban sprawl, informal land use or self-planned growth, environmental degradation, or disconnected housing and new town developments; each of them coming with a particular set of challenges for urban mobility to address or react to. Lastly, the majority of urban growth will not happen in the already highly urbanised and developed regions of the world, but continue in Asia and Africa. This extreme localisation results from comparatively low-urbanisation levels and high-urbanisation rates, continuous natural as well as rural–urban migration-inflicted population growth, urban economic development, and transition from leading rural to industrial and service-oriented economies (Hoornweg and Pope 2017).

The exploration of urban mobility futures is thus justified by urbanisation itself, challenged by resulting spatial and societal developments, whilst requiring a response which takes globally divergent patterns into consideration.

Glocalisation

The second global trend, directly linked to urbanisation, is glocalisation. Glocalisation is a combination of the term *globalisation,* referring to the increasingly global markets, value chains, societies, networks, and interconnectivities, as well as the seemingly opposing notion of localisation. It describes a dualistic trend of both increasing global and local networks across domains. Globalisation has shaped today's world, primarily resulting from developments of the last two centuries, enabled by new modes of transport for people and goods and advances in telecommunications, amongst others. At the same time, trends towards localisation arise. Jeremy Rifkin's perspective on the next industrialisation revolution focuses on the trends towards decentralisation of energy production, the resulting uprooting of traditional centralistic power distributions, and the arising opportunity of more localised social networks, as well as pro-consumption patterns (Rifkin 2011).

Further, increasing concerns arise from the negative long-term impacts of anonymity and disconnection to the immediate social and environmental surroundings in urban areas, accelerated and enabled through increasing digitalisation and widespread uprooting from traditional community networks. Rifkin (2011) provides the example of disconnection to the natural habitat when growing up in urban areas and its assumingly detrimental impacts on humans' capacity for empathy. A multitude of studies set out to quantify the impacts of urbanisation and urban life on mental health. Whilst a negative impact on depression and stress levels is agreed on, the search for impacting factors goes in different directions. However, a relationship between mental health and the level of social capital, resident continuity, and provision of places for interaction and exchange, seems to prevail.

Looking back at the evolution of species, applying the concept of Mutual Aid introduced by Kropotkin (1902) as a supplementary or rather contrary concept to Darwin's survival of the fittest, can add to the localisation discourse. Kropotkin studied the evolution of various types of animals and human tribes and found patterns of community behaviours such as mutual aid as a significant contributor to the evolution and prosperity of the group. Starting in the very early history, he traces developments thereof surviving as important factors until the urban age, when trade communities or neighbourhoods continued to support each other and contributed to their community. The widespread urbanisation, migration, relocation, and expanding reach through mobility, as well as accompanying consequences such as geographically distant networks to family and friends or local anonymity, have resulted in minimal local community structures in urban areas (most notably, Appleyard 1982; Jacobs 1961).

Today, many projects, initiatives, and trends can be found which are intending to reverse this trend, with results thereof constituting an often-found reference in existing urban future scenarios. These include, amongst others, human-centric urbanism (building on Jane Jacob's and Jan Gehl's work), placemaking, the 15-minute-city, child-friendly or inclusive urbanism, know-your-neighbour initiatives, or platforms for support or local trading networks (Jacobs 1961; Gehl 2011; Gehl and

Svarre 2013; Allam et al. 2022). Despite of a lack of a systemic understanding and diverse responses, the dual role of global and local networks has defined the growth of urban societies and will continue to do so, resulting in varying drivers and constraints for urban mobility futures.

Urban Areas as Agglomeration Economics

As urban areas are, in many cases, the engines of the global economy, the economic dimension is as complex as the city itself. Most importantly, if sustainable urban development allows for or is accompanied by economic prosperity, its implementation and scaling up is more likely, hence economically sustainable. Different approaches for this exist, most notable the circular city, regenerative city, or green growth applied to cities. Another commonly related concept is that of smart cities (or smart sustainable cities), which will be discussed further in the following section. The link between mobility and economic development is paramount and can be simplified by access to opportunity. Or in other words, how many people have access to what jobs? How many companies have access to how many suitable employees? Or how many companies of the same field compete on the one hand whilst sharing knowledge (e.g., through employee mobility) or joint suppliers. This brings us to the interrelation of how cities are built and planned and how it affects mobility of its inhabitants.

Urban Form: Sprawl vs. Compactness

The first determining element for urban mobility is urban form which can be described as the built embodiment of the urban society, spanning macro-, meso-, and microscale (city, settlement/neighbourhood, building). Urban form is constituted of different layers, including the street networks, built environment, and land use/division (Pont and Haupt 2009; Oliveira 2016; Hillier 2009). The urban level (macro scale) includes the demarcation of the urban agglomeration and is necessary to understand larger interrelations, e.g., the accessibility to the economic centres or differences between core and peripheral areas. The community level (including both meso- and microscale) considers the built environment and includes, amongst others, built density, space allocations, proximities, or the density of street intersections. Various interrelations between urban form and sustainability have been studied, stating, for example, that smaller, denser, and more interconnected cities are more sustainable (cf. Adolphe 2001; Oliveira et al. 2014; Jabareen 2006; Fragkias et al. 2013; Dave 2010; Louf and Barthelemy 2014).

An often-quoted reference for the large-scale relationship between density and transport-related energy consumptions is the Newman and Kenworthy hyperbola (Box 2.3) which shows a strong correlation, as well as a pattern of urban areas from different global regions.

In the mobility-specific context, Le Néchet (2012) points out that density, sprawl, and polycentricity can '*enrich the study of mobility patterns*' and that the transport-related per capita energy consumption is partly related to elements of urban form. He found that '*energy consumption is larger in a rich, motorised, sprawled, diffused and polycentric city*' whilst the contrary compact city seems to have a higher fatality rate. Without defining the right level of density or spatial characteristics, Le Néchet (2012) points out the importance to study the complex relationship of urban form and mobility and consider it for imagining and planning future urban mobility.

Box 2.3 The Newman and Kenworthy Hyperbola [Facts]

Urban density and transport-related energy consumption. The higher the density, the lower the transport-related energy consumption (adapted from Lefèvre and Mainguy 2009, based on data from UNEP 2008, original source: Newman and Kenworthy 1989).

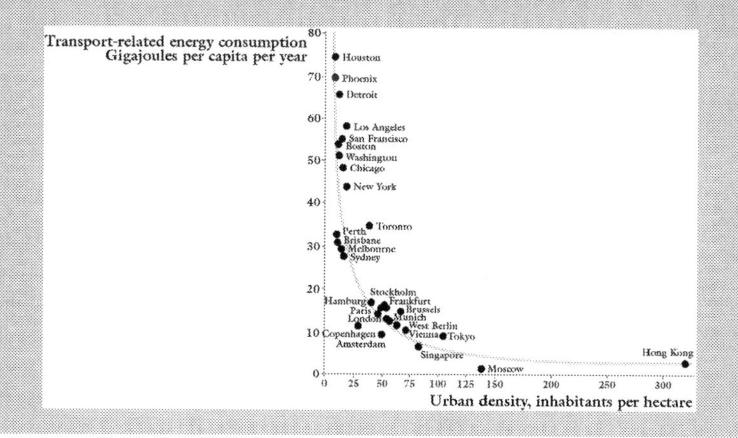

Urban form is also intertwined with mobility on a smaller scale. The density within a certain area defines the number of inhabitants and the related services or functions. Various decisions are made based on the density of an area, both from the public and private side. For example, the number, location, and size of healthcare services, schools, supermarkets, or food franchises is oftentimes based on the population in the catchment area thereof. Furthermore, higher density increases the number of places which can be reached within a certain timeframe. However, monofunctional density (e.g., a business centre or large-scale housing project) comes with various challenges. Hence, high density must be combined with a multifunctional mixed-use neighbourhood to fulfil its primary functions. This relationship has been integral to most historic cities and was re-introduced under various names over the last decades. Most recently, the term of the 15-minute-city has gained global attention due to its re-conceptualisation and advocating by Sorbonne Professor Carlos Moreno, and its integration into the re-election campaign of Paris' Mayor Anne Hidalgo (Allam et al. 2022).

Many more determinants of urban form and mobility have been studied which inclusion is outside the scope of this book. These include, for example, the optimal block size, number of network nodes, complexity or organicity of urban environs, active facades, social security due to mixed-use, as well as lightening and air movements, all with impacts on, for example, a higher likelihood of using, visiting, or passing through a space, hence impacting mobility patterns. Depending on the spatial granularity and case studies of the forthcoming scenario development, the considerations listed above will need to be studied further.

Transit-Oriented Development (TOD) and Multimodality

Lastly, and in many ways connected to the discussion on density and mixed-use, the concepts of TOD and multi-, or intermodality shall be outlined. TOD means '*integrated urban places designed to bring people, activities, buildings, and public space together, with easy walking and cycling connection between them and near-excellent transit service to the rest of the city*' (ITDP 2020). Combining a variety of urban design and planning principles, it describes the approach to develop urban areas depending on the proximity to mass transit hubs to increase the reach and impact of public transport. In the optimal case, the transit hubs combine a variety of modes (e.g., train, metro, bus). The density shall be the highest around the transit hub and slowly decreasing thereafter. The built environment shall be constituted of varying mixed-use zones, structured by their proximity and target group. The location of specific functions such as public amenities is often integrated to reduce the need for additional stops. The concept is widely applied around the world, mostly shaping the development of rapidly growing urban areas without or with little formalised public transport. However, their applicability and use in areas such as the Île-de-France remains the same, and hence is also one of the underlying principles of the Grand Paris Express.

Directly linked to the concept of TOD is that of intermodality and multimodality, both referring to the combination of different modes of mobility. Multimodality refers to the use of multiple modes of mobility, either for the same or different trips. Intermodality further adds the dimension of the seamless integration of modes of mobility within the same trip. This concept comes along with different elements. First and fundamental is the physical intermodality hub (such as a TOD) which must effectively combine various modes of mobility. This can, for example, include regional trains, metro, a bus station, as well as bicycle parking/rental, possible connections to micromobility, as well as ride-hailing/-sharing services and sufficient park or waiting areas, all easily accessible, connected, and clearly indicated. Secondly, the different mobility providers need to align their services and allow for seamless transitions. This includes, but is not limited to, the alignment of schedules, shared associated services (e.g., ticket sale/information) as well as collaboration in the payment system (e.g., through shared use of payment platforms, transportation cards, or abonnements). The third and last element addresses some of the above through the use of digital platforms which combine different services, starting from trip planning and going up to payment of different mobility providers. Whilst the latter diverges from urban form, it introduces the next chapter of digitalisation and will be extended on further.

2.3 Technological Trends

Aside from the trends and uncertainties in society and urban governance, the most discussed field in media and politics is technological innovation. This refers both to the actual technologies to move people from one place to another, as well as the advances in planning and design approaches. This section discusses those that we consider to be the most impactful today and in the years to come, without attempting to cover everything. Whilst many of the technological developments are seen as an integral part for a sustainable transition towards net zero emissions, we refer back to the Avoid-Shift-Improve framework that calls for avoiding trips and shifting to sustainable modes in priority, before focusing on the efficiency improvement. Further, many technological solutions are dependent on rare materials and other limited components along the value chain, such as semiconductors or microchips, which might depend on the region and geopolitical developments, put a significant strain on them. Lastly, the Jevons paradox highlights the challenge that each improvement in, for example, energy efficiency and thus decreased demand is compensated in increased demand. A related example includes that most gains that have been gained through electric vehicles over the last years have been nullified by the exploding market share of SUV with a much higher energy demand. With these limitations in mind, the following sections discuss some key technological development, starting with the structural and physical ones, and ending with the methodological and operations ones.

Electric Vehicles

From the technologies that define discussions around innovation and decarbonisation, none is as dominant as electric vehicles (EVs). A market rise in the Western world is mostly discussed since Elon Musk's Teslas. However, some of the earliest vehicles were EVs and mostly for smaller vehicles, EVs had always some niche market share. Much more so—until today—in Asia and China in particular. Nevertheless, the current variety of models, the customer acceptability, and the scale has developed rapidly over a short period of time (IEA 2023). Amongst EVs, we can distinguish further between Battery Electric Vehicles (BEVs) which are only relying on batteries, Hybrid Electric Vehicles (HEV), and Plug-in Hybrid Electric Vehicle (PHEV). HEVs are generating and storing energy whilst braking and support the Internal Combustion Engine (ICE) to reduce emissions and fuel consumption. PHEVs have an added possibility to charge the batteries also externally, increasing the autonomy (i.e., range) of only battery-powered use. Whilst each has again advantages, in reality, the use of the electricity-powered functionality is rather low.

Thus, HEV and PHEV can result in higher consumption than a traditional ICE vehicle due to the added weight resulting from batteries and supplementary technology.

Focusing back on the BEV, the dialogue today circles mostly around individual personal or logistics vehicles. However, the biggest market total and growth share are made of from 2- and 3-wheelers, mostly in Asia and increasingly Africa (ITF 2021; IEA 2023). Whilst it is impossible to discuss all challenges and potential around EVs in this scope, we briefly focus on those that dominate current research.

Decrease of GHG emissions: Foremost, EVs shall reduce the use of fossil fuels and the emissions of Greenhouse Gases (GHGs). In most cases, EVs are significantly outperforming their ICE alternatives when comparing emissions per passenger kilometre, including public transport in case of Diesel-fuelled busses or heavy rail transport with low-occupancy rates. However, the energy still must be produced, thus shifting the emissions to the electricity production and transport. EVs environmental sustainability is largely linked to the energy mix of the country or region of use. In contexts with coal-heavy energy mixes (e.g., Poland, Germany), EVs easily emit more GHG than relatively new ICE cars. In countries with high levels of renewable or nuclear power (e.g., Norway, France), EVs are outperforming fuel-based cars continuously. Whilst this challenge will most likely vanish over time due to a more sustainable overall electricity network energy mix, it still requires a context-specific lens.

Decrease of local PM2.5 and noise pollution: Pollution and emissions are often treated together, yet constitute two related but independent challenges. GHG emissions must be reduced due to their negative impact on the atmosphere and thus global impacts. Local pollution is mostly a problem of older cars, highly impacted by density, topography, and climate. Wind, rain, low densities, and lower temperatures are usually helping to quickly distribute the local pollution whilst the inverse can contain it for a long time at street level. Highly polluted cities such as New Delhi or Beijing are most impacted when there is little wind, high-traffic volumes, dry and hot periods. This leads to moments when air and ground traffic and industry must be halted to reduce levels immediately. And this reduction is just to make it slightly less bad—they remain nearly constantly at highly critical values. Currently, it is assumed that up to 1.4 million people die every year because of air pollution. Most of these problems significantly decrease with newer ICE cars and nearly disappear with

EVs. Noise pollution reduces significantly. However, one main challenge remains: EVs are usually heavier due to the batteries, which combined with an overall trend to larger cars and SUVs, lead to increasing tire abrasion with significant impact on local air pollution. No technological fix exists for this yet, even if tire producers try to reduce the impacts. Thus, EVs are much better for local pollution but have some unresolved challenges usually ignored.

Range and charging infrastructure: A key argument against EVs in the beginning was the range. This has multiplied since but remains a problem, mostly for inter-city or longer trips (Baltazar et al. 2022). Further, reliability on the stated range remains limited as older batteries or extreme temperatures impact significantly the possible ranges. The challenge is to navigate between an acceptable range and battery size: Every increase of battery capacity comes with an added weight, again reducing the capacity. The other challenge is on the user-side: Every market EV today addresses the large majority of the usual personal use case. However, customer choices are also made based on factors such as the distances moved once per year during the holidays. This brings us to the next point of recharging. EVs recharging remains challenging. Some issues, such as interoperability between EVs and charging station providers, have been addressed largely via regulated and voluntary standardisation. Yet, even with high-powered chargers, the waiting times are much longer than for fuel-based cars. In daily use, this rarely constitutes a problem if charging stations are available at work, in the neighbourhood, or a home-charger is installed. However, for long or full day trips, charging remains difficult due to the time as well as the charging infrastructure. The latter is well developed in some countries and densely populated areas but still lacks to varying degrees in most rural areas. Nevertheless, the US as well as some Asian and European countries, invest heavily in the deployment. The range and charging infrastructure density constitutes primarily a problem for long range logistics whilst most urban use cases have been shown to be possible without significant restrictions. For this, different solutions including induction recharging at waiting points or overhead charging infrastructure on parts of highways are explored yet so far only limited to trucks and in some cases busses and early pilots. The scale, pace, and geographical distribution of EV growth is thus largely attributed to the development of the charging infrastructure and the technological, range-increasing innovation.

Box 2.4 Battery Degradation [Facts]

Battery degradation is a fundamental concern for EVs, and it is particularly relevant when considering the integration of SAEVs into the grid. Batteries in EVs undergo degradation due to various factors. Cycling, which involves charging and discharging the battery during regular use, contributes to wear and reduces the battery's capacity over time. Calendar ageing, a process where batteries degrade even when not in use due to chemical reactions, is another significant factor. Extreme temperatures, both hot and cold, can accelerate battery degradation, with high temperatures causing faster internal chemistry degradation and cold temperatures reducing reaction efficiency.

Furthermore, the way EVs are used influences degradation rates. Frequent rapid charging and heavy loads, such as towing, can expedite battery wear. Battery degradation directly affects the driving range of EVs, diminishing the distance they can travel on a single charge.

Researchers like Wang et al. (2016) have addressed battery degradation by quantifying it, considering factors like cycling and calendar ageing, especially in situations where EVs participate in Vehicle-to-Grid services, meaning the battery charge can be fed back into the grid when needed. Bishop et al. (2013) expressed concerns about Vehicle-to-Grid services accelerating battery wear and suggested reducing battery capacity as a mitigation strategy, although this may not be practical for use cases such as Shared Automated Electric Vehicles (SAEVs) due to range restrictions. Lee et al. (2020) highlight the importance of optimising battery capacity in relation to fleet size, considering factors like cost and charging station wait times. Randall (2016) provides an outlook, suggesting that advancements in battery technology beyond 2030 could reduce reliance on certain materials, making batteries lighter and cheaper.

Whilst battery degradation remains a challenge, ongoing advancements in battery technology and management systems are gradually improving the longevity and performance of electric vehicle batteries. However, it is essential to be aware of the issue and best practices for battery care, such as avoiding extreme temperatures, using appropriate charging methods, and adhering to manufacturer-recommended maintenance schedules to maximise the lifespan of their EV's battery.

Electricity grid pressure and relief: Assuming that the charging infrastructure is developed, the next challenge is the demand and the locations. An increase of electricity consumption in residential, dispersed, or already highly used areas puts a significant pressure on existing networks. This is further aggravated by increasing decentralised electricity grid feeding from overhead renewable energy production from households' solar panels. One challenge is the distribution on local and regional levels, another is the storage and temporal distribution. Solutions include the rapid extension of critical infrastructures, decentralised and centralised storage capacity increase, innovative pricing schemes to distribute supply and demand over the day, and exploratory concepts such as vehicle-to-grid solutions which permit to use EVs connected to the grid to act as temporary energy storages. Many solutions already exist or are in development. However, their scale and pace will—once again—impact the possible development of the EV market.

Furthermore, the rapid increase in EVs penetration is expected to impose a significant additional load on electric power grids, leading to potential security risks for the power systems (Liu et al. 2011). The uncoordinated charging of EVs, which often occurs when they are plugged in without a planned charging control, can cause grid congestion during low-cost periods, resulting in various impacts on the power system, such as thermal limit violations due to the overload of the feeders and the transformer, voltage profile degradation, and increased operational costs (Liu et al. 2011). To address the EV charging problem and mitigate these challenges, intelligent management systems derived from proper management and optimisation frameworks are necessary. These systems should efficiently coordinate the charging process amongst EVs, aiming to avoid grid overload whilst ensuring cost-effectiveness for EV users and meeting their transportation and charging requirements.

Different approaches have been proposed in the literature to model and optimise the coordinated charging of EVs whilst avoiding grid congestion and overload. Amongst these approaches, optimisation-based methods have been explored. These methods consider various factors such as EV owners' mobility patterns, which may lead to similar charging patterns causing congestion during peak demand periods (Cao et al. 2012). Additionally, charging energy cost can influence the charging behaviour of EVs (Karfopoulos and Hatziargyriou 2012). However, some optimisation-based papers adopt a centralised approach to control EV charging, which can be efficient for managing a limited number of EVs but faces challenges when dealing with a large fleet due to the need for significant computational resources to process local information from a central point, including battery state of charge and desired charging period.

An effective and promising approach for the coordinated charging problem is to treat each EV as an intelligent system capable of autonomously choosing its own charging schedule. This approach typically models EVs as agents in a multi-agent system. Multi-agent systems have found applications in various domains, including transportation, logistics, manufacturing, power systems, and smart grids (Xie et al. 2012; Singh et al. 2017). In a multi-agent system, agents interact by sharing knowledge and negotiating with each other, pursuing either their individual interests or some global goal collaboratively. The interactions occur within a dynamic, unpredictable, and open environment. At the core of a multi-agent system, agent interactions play a crucial role in shaping overall behaviour.

As the interest in artificial intelligence and machine learning continues to rise, researchers have explored the application of machine learning approaches to handle interactive dynamical systems in complex environments (Duan et al. 2016). Reinforcement learning, a branch of artificial intelligence, has shown great promise in handling demanding tasks such as autonomous driving and decision-making problems (Mnih et al. 2015). Multi-agent reinforcement learning systems have shown the potential to overcome the limitations of other optimisation-based approaches and have demonstrated efficiency in addressing the coordinated EV charging problem, even with a large fleet. A more detailed discussion of this challenge and possible remedies follows in the data-driven design and decision support sub-chapter.

Fire: A very short point on the risk of battery fire as it is frequently reappearing. Despite its overreporting, ICE vehicles remain 60 times more likely to catch fire (hybrid twice as likely as ICE) and thus should not constitute a factual barrier to EV adoption—even if its perceived risk can continue to impact purchase behaviour.[3] On the other hand, other factors might influence the comparison and temper with its reliability. For example, the average age of EVs is lower and many of them are high-quality vehicles compared to average ICE vehicles. Nevertheless, EVs do currently not appear to constitute a higher risk compared to other vehicles.

Raw materials/battery recycling: The core element of EVs is the energy storage in batteries. Whilst some old and more polluting ones remain to be used for 2- and 3-wheelers, most batteries are lithium batteries. The value chain, starting from the resource extraction up to battery production, constitutes another barrier to battery supply and thus the pace and price of upscaling. Three challenges already exist or are expected to have a significant impact on this process: (1) The raw materials used for batteries need to be extracted in large quantities, requiring increased investment in mining infrastructure. New, alternative, and optimised technological innovations might decrease the pressure partially. (2) The growth of battery production capacity does not match predicted demand yet. With lots of investments and public subsidies, this is highly discussed but not resolved yet. (3) Geopolitics: Like oil, gas, or uranium in past and present, raw material resources are not distributed equally around the globe but found in few specific locations, leading to inherent power inequality and conflicts (Rifkis 2011). The race which country has access to whose resources has started, dominated by Asia, the US, and Europe. The Russia-Ukraine war that started in 2022, the 2023 US Inflation Reduction Act, EU responses and movements towards more protectionism and de-risking, as well as ongoing critical global relations between China, US, and EU, are some of the driving factors. The conflicts and collaborations resulting from the new scramble for (mostly) Africa, this time for other raw materials, will define who can advance EVs how fast.

Hydrogen and e-Fuel Vehicles

Less often discussed and—at least for now—less relevant in the urban passenger mobility context is hydrogen. As with most solutions, we can usually not say that one is better than the other, but all have advantages and disadvantages that make them interesting for specific use cases. So

[3] See https://www.autoinsuranceez.com/gas-vs-electric-car-fires/ [accessed August 2023].

far, hydrogen is considered mostly for long-distance trips in areas with less recharging infrastructure, heavy freight transportation, and use cases that require fast recharging.

Nevertheless, at the current stage, there is very limited availability of hydrogen refuelling infrastructure in urban areas. Further, even if hydrogen use can be considered as a sustainable energy source, its production and transport require significant energy. A set of colours is used to describe the differences. Grey, blue, and turquoise hydrogen is created via natural gas with different technologies and black hydrogen via coal. Thus, all of them come with significant emissions in the early process. Pink hydrogen uses nuclear power, yellow solar power, and green general renewable energies. Thus, all of them have reduced emissions. The most promising path is currently described as creating hydrogen in places where plenty renewable energy can be created cheaply and where all other resources are available, e.g., Namibia. This risks to result again in a centralised approach resulting in high-transport volumes from areas of production to use. Due to the early stage of development, a certain hesitation remains currently a barrier for investment and scaling as it is assumed that the unit costs will decrease soon, making any earlier investment a comparative loss. Thus, some state actors are currently primary project supporters to bridge this finance gap.

On a more local scale, hydrogen fuel cell vehicles also remain to have higher upfront costs compared to conventional vehicles, making them less affordable. The local storage and transportation of hydrogen remains challenging due to its low density and requirements of specialised tanks and infrastructure. Regulatory and safety issues are neither standardised nor clear. The highest concern remains flammability and explosion risks which require additional operational safety measures, leading, for example, to the interdiction of using hydrogen vehicles in some tunnels and thus creating significant use case limitations.

Thus, whilst hydrogen-powered technologies come with several advantages, their readiness level lacks far behind electric vehicles, and various open questions remain, revolving around the energy use for production, large-scale transport, distribution networks, safety, and regulation. Yet, continuous research, pilots, and investment render a tipping point in the near future very plausible.

Another fuel type gaining traction in recent years is the use of e-fuels, short for electro fuels or synthetic fuels. They are referred to as a promising alternative to traditional fossil fuels. Unlike conventional fuels derived from petroleum, e-fuels shall be produced using renewable energy sources and captured carbon dioxide (CO_2). The production involves converting CO_2, either directly captured from the air or from industrial emissions, into synthetic hydrocarbon fuels such as methanol, ethanol, or synthetic diesel. A key advantage is the compatibility with existing ICE and fuel distribution infrastructure. They can be used in conventional vehicles without requiring major modifications. This makes e-fuels a potential solution for decarbonising existing fleets and substituting EVs or hydrogen in areas where recharging infrastructure is unviable.

Another advantage of e-fuels is the theoretical carbon neutrality or even carbon negativity. By capturing CO_2 during production, e-fuels could offset emissions generated during their use, resulting in a net-zero or negative carbon footprint. However, the reality is far from this status, despite strong political pushes, mostly by car producing countries such as Germany. The process still requires significant amount of energy and scaling up production to meet potential global demand does not seem viable (yet). Additionally, the production costs and energy demand of e-fuels are currently significantly higher compared to conventional fossil fuels or other alternatives.

Furthermore, the overall efficiency of e-fuel production, including the energy losses during conversion processes, are not at a stage yet where benefits outweigh the energy inputs. Despite these challenges, e-fuels have a potential which is—however—difficult to assess at this stage.

Automated Vehicles

Aside from different energy sources of vehicles, a highly discussed topic is that of AVs. Whilst not implicit, AVs are usually BEV, thus similar constraints as above apply. A first important factor is what AV stands for. Traditionally, it referred to autonomous vehicles, implying that the vehicle is fully independent of a driver. Recently, automated or automatised start to dominate. This comes with a lower level of expectation and thus also potential responsibility of the technology. The standard scale of degree of autonomy can help, which starts at level 0 with no driving automation, over 1, driver assistance, 2, partial driving automation, 3, conditional driving automation, 4, high driving automation, and 5, full driving automation (SAE J3016[4]). Thus, whilst AVs might be used to refer to autonomous, market solutions are mostly between level 2 and 3, whilst pilot projects sometimes go up to 5, yet mostly with continuous supervision. Aside from its technological fascination, AVs are primarily promoted for five reasons: (1) People may create accidents, drive drunk or aggressively whilst AVs shall achieve higher safety standards. (2) Human reaction time and non-optimised driving requires longer safety distances on roads. Through V2V (vehicle-to-vehicle) communication, AV fleets could result in high-speed fleet management with much higher vehicle throughputs on existing infrastructure. (3) Along the same lines, AVs can better adapt speeds to conditions, including due to knowledge of the traffic management, e.g., via V2I (vehicle-to-infrastructure) communication. This can lead to higher efficiency and decreased energy demand. (4) Human drivers are currently usually the highest cost of transport solutions, in case of public transport, for example, prioritising large capacity vehicles. AVs could permit more diverse public transport and on-demand fleets with a more adapted and flexible transportation system. (5) Finally, primarily shared AVs permit new use cases, such as connecting rural areas with mass transit or increasing affordable accessibility for disabled or the elderly. Whilst just representing some of the key potentials, the current barriers are technological primarily regarding the ability to perform safely in mixed (human and AVs) and changing inner-urban contexts, regulatory constraints regarding V2V communication between operators and data privacy, as well as legal ones. Open questions are, who is held responsible for accidents or how decisions without optimal solutions are made. For the latter, the Moral Machine by MIT best demonstrates the challenges by crowd-sourcing moral decisions in difficult situations AVs might face.[5]

Concluding, many potentials and barriers remain, but those can be expected to be resolved to different degrees in the near future. Regarding the impact on the overall urban system, key determinators are if AVs would be primarily shared or individual, as larger shuttle or small vehicles, and if they are dominated in low-density areas where they can best supplement inefficient public transportation or if they might add increased pressure in already highly densified areas. For a more complete discussion of potential directions, we can highly recommend the set of scenarios by

[4] https://www.sae.org/standards/content/j3016_202104/ [accessed August 2023].

[5] https://www.moralmachine.net [accessed August 2023].

Townsend (2014), or a more creative lens from the Prospective Atlas of the Robomobile Planet (Robomobile Life Workshop 2020).

Technology in Cities—Urban Mobility Ecosystems in Reconfiguration

For the better part of the last century, technology became a component of cities that gained continuously traction. Driven by widespread digitalisation, urban mobility innovations are often driven by objectives such as cost optimisation and revenue maximisation, with secondary objectives of resolving issues such as congestion or pollution. Specific motivations of the mobility context are meeting the needs of a society that is increasingly mobile and more aware of environmental issues (Le Breton 2019). However, technological innovations also play a role in mitigating the externalities produced by mobility in cities and in creating alternative, less energy- and space-consuming options that are more in line with current climate issues. The modernisation of urban mobility systems is trying to keep up with the evolution of individual mobility practices, facing the urgency of climate and environmental issues and evolving economic and development systems.

Information and Communication Technologies (ICT) play a particularly important role in integrating innovations within transportation systems, for example, by integrating real-time digital data into route calculations. Since the arrival of smartphones, there has been a new production and exploitation of location data. Payment optimisation has also played an important role in improving transportation services and creating new ones (Aguilera and Boutueil 2019). The digital technological innovation in transportation, made possible by ICT innovation, gives historical mobility actors tools to act across levers: pricing and ticketing, traveller information, intermodal journeys, security, practice analysis, management, and maintenance, to name a few (Mouly-Aigrot et al. 2016). These technological innovations also provide access to advantages such as optimisation of operating conditions, cost reduction through better planning, and the participation of the user as an active actor in reporting traffic or infrastructure problems.

The innovations in transportation systems have developed through improvements of all kinds, whether they are technological improvements or managerial and interaction processes. Innovations have also made it possible to optimise the functioning of mobility systems and facilitate their organisation (Docherty et al. 2018; Bonilla and Carreon-Sosa 2020).

Innovation in the transportation sector is also related to societal innovations, as witnessed through the change of trends in the demand side of transportation.

Punctual innovations such as those in the telecommunications sector—high-end sensors in the smart city, global positioning systems (GPS), amongst others—are likely to help optimise logistics and freight, facilitate travel planning with private cars, simplify toll and other traffic fees payments, and allow for the improvement of public transportation to regain some of the space taken up by private cars in public space (National Academies of Sciences, Engineering, and Medicine 2016).

Regarding innovations in urban public transportation, the arrival of managerial innovations (Agarwal et al. 2018, p. 18; Paap and Katz 2004) is important for the collaboration objectives that trends like Mobility-as-a-Service (MaaS) pose (Agarwal et al. 2018, p. 18; Paap and Katz 2004; Mouly-Aigrot et al. 2016). An example to explain this is the logic of responding to the increase in demand with an increase in infrastructure supply (Paap and Katz 2004). However, transformations in the interactions of actors of the mobility ecosystem are necessary to achieve these objectives.

Platformisation

The way we currently access mobility services through our smartphones is different from the existing approaches ten years ago (Aguilera and Boutueil 2019). As mentioned earlier, we are witnessing the development of two new forces that shape social dynamics around mobility: the service economy (Rothenberg 2007) and technological innovation (Autio and Thomas 2014; Bower and Christensen 1995; Agarwal et al. 2018; Geels 2004). These two forces have changed the way public transport is organised. The increase in the number of shared mobility services in urban areas and the way these modes are organised are modifying mobility dynamics and making door-to-door travel more accessible on a global scale. One of the most frequent arguments for MaaS is environmental sustainability (Sarasini et al. 2017). It has indeed been mentioned that MaaS is supposed to reduce the number of private cars on the streets and encourage users to give up owning a car (Li et al. 2019; Smith 2020). However, to make MaaS a competitive system, MaaS solution providers must ensure a continuum of services to meet demand and match competing service quality levels.

However, major transformations are occurring and the shift is pointing towards innovation in transport demand management with specific innovations such as the opening and real-time access to traffic data (Aguilera and Boutueil 2019; König et al. 2016; National Academies of Sciences, Engineering, and Medicine 2016). A turning point in the conception of mobility is also taking place in connection with technological and managerial innovations: that of user-centred mobility. This new approach places users at the centre of the business model as consumers and producers of data (Lesteven et al. 2018, p. 141). In summary, a new value chain has been created. Value capture is based on the logic of access to seamless multimodal, intermodal, and door-to-door mobility.

Public authorities have valuable opportunities for optimisation of service quality through innovation in the areas of organisation, planning, security, funding, and access to public transport information systems (Mouly-Aigrot et al. 2016). The public sector has increasing opportunities to harness data collected by intelligent systems in 'smart cities' to better manage urban mobility (Aguilera and Boutueil 2019). One of the challenges is acceptance and success in collaborating with several actors who were not previously part of the transportation ecosystem but who have entered as stakeholders in the current mobility entrepreneurial sphere (Mouly-Aigrot et al. 2016).

The assimilation and integration of private sector technological innovations by the public sector could give the authorities responsible for organising public transport systems a lever against the technological disruptions currently emerging in the mobility ecosystem. Technological disruptions, as discussed by Paap and Katz (2004), are, in fact, the effects that certain technologies have on markets affected by technological innovation. This results in a process of loss of power of historical entrepreneurial actors when they compete with new actors without having adapted at the right time and with the right strategy.

New mobility services, including shared ones, appear to be the result of an evolution of the needs of urban users, motivated by 'socioeconomic, demographic, technological, and environmental reasons (…)' (Huré 2019). There is an increase in the number of shared mobility services, the number of companies involved, and the number of trips they offer. These services include carpooling, car-sharing, shared bicycles, on-demand shuttles, as well as micro-transit. In this category are traditional shared mobility services such as taxis in competition with new services (National Academies of Sciences, Engineering, and Medicine 2016; Aguilera and Boutueil 2019).

Around 2016, new mobility services arrived in cities and began increasingly to change the actor and regulatory landscape of urban mobility systems. Cities were being actively supplied by free-floating bike and scooter services. In the French case, some new market players have no direct link with historical transport actors, and others have started their service in cooperation with existing ones. In both cases, they find themselves in a new situation of competition for urban mobility demand (Mouly-Aigrot et al. 2016). Competition with new urban mobility operator-actors is now institutionalised in many places, strongly encouraged in European directives (EC 2011). This competition plays a key role in stimulating innovation and promoting improved service quality.

The cooperation between mobility actors (including new ones) is one of the main recommendations made by researchers to historical actors in the face of technical innovations in transport systems (Aguilera and Boutueil 2019; Bonnafous 1996; Smith 2020). This actor cooperation could ensure more integrated mobility and optimise value capture by public institutions such as AOMs.

In summary, the combination of ICT and recently developed shared modes has led to the emergence of new mobility services such as MaaS. However, shared mobility services face significant challenges. Therefore, if they plan to remain an attractive and competitive alternative in today's mobility landscape, they must overcome the challenges of collaboration between actors (Kamargianni et al. 2016). Another challenge is the modification of user behaviour towards more personalised travel (Ho et al. 2018). The disposition and willingness of users to adopt MaaS solutions for their travel could also play an important role in the acceptance of this technology as a competitive mobility alternative.

Technology in the Sustainable City

Sustainable city is an umbrella term to describe developments that make urban areas more environmentally friendly, socially inclusive, economically productive, and equal, and participatory and democratic. This contains considerations such as demographic change (ageing, migration, and mobility), social mixing and cohesion, identity, sense of place and pleasure, empowerment, participation and access, health and safety, social capital, as well as well-being, happiness and quality of life (Colantonio and Dixon 2009).

Social capital, well-being, and happiness are themes appearing today, when the social pillar of sustainability is addressed. Such themes complement more traditional topics such as social justice, poverty, or employment. Boosting happiness and well-being is also the preferred way towards a fair, happy, and low-carbon city for Montgomery (2013). For this author, a city is above all a 'shared project'. Examples of requirements of a fair city are: (1) safe streets, especially for children; (2) access for all to parks, shops, services, and healthy food (Montgomery 2013, p. 250).

The City of Paris[6] illustrates her dual vision combining technology and sustainability. It is engaging a strategic plan to develop a 'smart and sustainable' city, where the driving forces are technological solutions and open innovation. This development relies on the participation of Parisians and of the 'innovating ecosystem' (i.e., start-ups and companies).

According to the French Ministry of Ecology, Sustainable Development and Energy,[7] Intelligent Transport Systems rely on ICT to 'pave the way to new mobility services'. The corresponding National Strategy, called Mobility 2.0, was launched in 2014 on the basis of four main axes: (1) deployment of cooperative systems (connected vehicles); (2) development of open data for transportation; (3) development of multimodal travel planner; (4) set national consolidated priorities across stakeholders (industries, public actors, researchers). Despite a growing complexity in mobility flows, two factors are expected to foster progress. It concerns, on the one hand, open data in the transport domain to give access to information on multimodal displacements; on the other hand, new modelling techniques.

ICT is one of the five key research themes identified by Stead (2016) for sustainable urban mobility. The general framework of interest seeks to describe how technology shapes society. More specifically, consequences of ICT for sustainable urban transport are examined by scholars. It is interesting to investigate if ICT either stimulate more travel, substitute for travel with remote activities (e.g., teleworking), or modify travel in some ways. Many questions remain open, notably the balancing effect between substitutions or complementarity effects of ICT on transport (Stead 2016).

[6] http://www.paris.fr/services-et-infos-pratiques/innovation-et-recherche/ville-intelligente-et-durable [accessed August 2023].

[7] http://www.developpement-durable.gouv.fr/Mobilite-2-0-une-strategie.html [accessed August 2023].

Debnath et al. (2014) elaborate indicators of smartness to set a proper concept of a smart transport system. Based on data collected in 26 cities throughout the world, the authors defined 66 indicators for basic and advanced smart requirements. Basic requirements include collecting information through sensing and monitoring, processing information, acting and controlling, communicating between sensors. Advanced requirements involve predicting problems, healing situations, and preventing potential failures. The indicators are applied to evaluate private transport, public transport, and commercial or emergency transport. According to this set of indicators, London, Sydney, and Seattle appear amongst the top smart transport cities.

A highly discussed solution, responding to the earlier discussed trend of servitisation, is MaaS. It is often used as a term to describe an application or interface to access different mobility modes. Yet, MaaS is rather a concept that continues to evolve whilst having faced stagnation when it comes to showing the expected results and impacts on sustainability (see Box 2.5).

Box 2.5 Mobility-as-a-Service (MaaS)

Defining MaaS has been an interesting, complex process that has required the participation of each stakeholder involved in its creation, implementation, and deployment. Many conceptualisations for MaaS have developed over the last decade. MaaS has not evolved as a linear process, instead, it has complexified as a network of interwoven processes integrated thanks to ICTs (Streeting et al. 2017). For example, MaaS can be considered as a development encompassing new methods for planning and managing mobility (Mulley 2017). Other authors consider MaaS as a new distribution organisation of all the means of transport available in central platforms that would allow us to be a click away from optimised options to reach our destination (Hietanen 2014). Digital integrated mobility solutions such as MaaS have been in the limelight of roadmaps towards sustainability, with high expectations but little results (Hensher and Hietanen 2023). Since

MaaS appeared as a 'new concept' in 2014, it has been through periods of hype and disillusionment as described by Hans Arby for Futura mobility (Brown 2021). Both, physical and digital mobility services have encountered multiple barriers to stay economically viable and to overcome organisational hurdles. As observed by Offner, a researcher and expert in the field of urban mobility (interviewed by Reyes Madrigal in June 2023), one of the main learnings of MaaS so far is the awareness that the main obstacle to the development of the service layer within mobility systems remains the lack of coordination amongst actors and stakeholders involved. Five levels of integration have found consensus when speaking of MaaS, from level 0 reflecting no integration between mobility services, to level 4, where not only the available mobility services are integrated but also different payment options and services bundles, as well as societal goals through public policy (Sochor et al. 2018).

Here, we see MaaS as a people-centred technological and organisational innovation for mobility management. In this scope, MaaS aims to reduce negative transport externalities, mostly provoked by the excessive use of the private car by simplifying access to mobility for users whilst providing quality and useful information for their trips (Reyes Madrigal et al. 2023). The innovation in MaaS lies in the integration of a diversity of modes into a digital one-stop-shop, for example, a smartphone app or a website.

The use of ICT is ubiquitous in many cities worldwide and seems to carry many promises towards a sustainable urban life. This is illustrated, for instance, by the strategic plan for the City of Paris. However, the conjunction of sustainability and 'smartness' in cities deserves a closer look. Dealing with smart cities, Picon suggests that any form of technological determinism must be avoided (2015). About urban mobility, he judges that mobility is bound to keep a random character, despite any attempt of prediction (Picon 2015). Flipo et al. (2013) discuss the environmental impacts of digital technologies as enablers to reduce environmental impacts by, e.g., mitigating energy consumption. The benefits of ICTs are expected to be higher than the impacts caused by devices and data centres.[8] One should, however, not forget other related problems such as resource scarcity linked with rare metal use in ICT solutions, generation of electric waste, or rebound effects due to huge increase in data and data centres. For example, the progress made in energy consumption of household products is largely offset by the increase in number of devices and use. Therefore, whilst with many potentials, the environmental contribution of ICTs may only be relative and compensated in large parts by rebound effects (Flipo et al. 2013, p. 121). Kaack et al. (2022) discuss this for the context of artificial intelligence and machine learning which faces similar challenges. They propose a three-level framework to analyse the impacts of technological solutions, especially machine learning as computational heavy and increasingly used method in mobility and transport management and planning. At the first level, one must look at the computational demand. At this stage, a differentiation between the initial and demand-intensive training phase and the following use-phase is important. The second level is that of the application. For example, decreased car emissions due to improved traffic management. This is the one mostly talked about. On the third level, systemic impacts and rebound effects shall be looked at. For example, might faster vehicles lead to more demand or will lower energy demand for vehicles lead to increased purchases of larger vehicles such as SUVs? All of these are considerations crucial when considering technological innovation to achieve sustainability objectives.

The Smart Enough City

To summarise, we borrow the book title by Ben Green and propose the 'smart enough city' as an adequate way forward (Green 2020). This means using technology wherever and whenever, it can contribute to the sustainable objectives of urban development. On the other hand, not developing new solutions on digitising services for the sake of innovation. In the context, we can find plenty innovative example of smart enough urban innovation. For example, the use and real-time access of mobility information makes more complex routes and the combination of mobility modes possible. Geolocalisation and remote access permits the sharing of

[8] Causing about 2% of total CO_2 emissions in the EU and 10–15% of the total electric consumption (Flipo et al. 2013, p. 20).

vehicles of all kinds. Decentralised sensors for traffic volume, noise pollution, or air quality accessible to citizens empower the urban society and permit to hold the local governments accountable. Finally, a large set of open source and accessible tools and methods, combined with an ever-increasing richness of urban data permits to advance research, planning, and design and creates accessibility to methods that have been limited to few principal stakeholders in the past.

References

Adolphe, L. (2001) A simplified model of urban morphology: application to an analysis of the environmental performance of cities. Environment and Planning B: Planning and Design, Vol. 28, pp. 183–200. https://doi.org/10.1068/b2631

Agarwal, O.P., Zimmerman, S., and Kumar, A. (2018). *Emerging paradigms in urban mobility. Planning, financing and management*. Elsevier. ISBN: 9780128114353.

Aguilera, A., and Boutueil, V. (2019) *Urban mobility and the smartphone*. Elsevier. https://doi.org/10.1016/C2016-0-03595-7.

Allam, Z., Bibri, S., Chabaud, D., and Moreno, C. (2022) The theoretical, practical, and technological foundations of the 15-minute city model: Proximity and its environmental, social and economic benefits for sustainability. *Energies*, Vol. 15. https://doi.org/10.3390/en15166042.

Appleyard, D. (1982) *Livable streets*. San Francisco: University of California Press.

Autio, E., and Thomas, L. (2014) Innovation ecosystems. In: Dodgson, M., Gann, D.M., Phillips, N. (eds.), *The Oxford handbook of innovation management*. Oxford University Press, pp. 204–288.

Babiker, M., Bazaz, A., Bertoldi, P., Creutzig, F., De Coninck, H., De Kleijne, K., Dhakal, S., Haldar, S., Jiang, K., Kılkış, Ş., Klaus, I., Krishnaswamy, J., Lwasa, S., Niamir, L., Pathak, M., Pereira, J.P., Revi, A., Roy, J., Seto, K.C., Singh, C., Some, S., Steg, L., and Ürge-Vorsatz, D. (2022) *What the latest science on climate change mitigation means for cities and urban areas*. Indian Institute for Human Settlements. https://doi.org/10.24943/SUPSV310.2022.

Baltazar, J., Vallet, and Garcia, J. (2022) A model for long-distance mobility with battery electric vehicles: A multi-perspective analysis. *Procedia CIRP*, Vol. 109, pp. 334–339. https://doi.org/10.1016/j.procir.2022.05.259.

Bevan, L.D. (2022) The ambiguities of uncertainty: A review of uncertainty frameworks relevant to the assessment of environmental change. *Futures*, Vol. 137, p. 102919. https://doi.org/10.1016/j.futures.2022.102919.

Bishop, J.D., Axon, C.J., Bonilla, D., Tran, M., Banister, D., and McCulloch, M.D. (2013) Evaluating the impact of V2G services on the degradation of batteries in PHEV and EV. *Applied Energy*, Vol. 111, pp. 206–218.

Bongardt, D., Stiller, L., Swart, A., et al. (2019) Sustainable urban transport: Avoid-shift-improve (ASI).

Bonilla, R., and Carreon-Sosa, R. (2020). Transport and sustainability. In *International encyclopedia of human geography*, 2nd ed., Vol. 13. https://doi.org/10.1016/B978-0-08-102295-5.10150-7.

Bonnafous, A. (1996). Le système des transports urbains. *Économie et statistique*, Vol. 294/1, pp. 99–108. https://doi.org/10.3406/estat.1996.6087.

Bower, J.L. and Christensen, C.M. (1995) Disruptive innovation: Catching the wave. *Harvard Business Review*, Vol. 73/1, pp. 43–45.

Brown, L. (2021) Hans Arby: MaaS—mars 2021. Futura-Mobility. https://futuramobility.org/fr/hans-arby-maas-mars-2021/ [accessed August 2023].

Cao, T. et al. (2012) An Optimized EV Charging Model Considering TOU Price and SOC Curve. IEEE Transactions on Smart Grid, Vol. 3/1, pp. 388–393. https://doi.org/10.1109/TSG.2011.2159630

Colantonio, A., and Dixon, T. (2009) Measuring socially sustainable urban regeneration in Europe, Oxford Institute for Sustainable Development (OISD), School of the Built Environment, Oxford Brookes University, 2009.

Creutzig, F. (2016) Evolving narratives of low-carbon futures in transportation. *Transport Reviews*, Vol. 36/3, pp. 341–360. https://doi.org/10.1080/01441647.2015.1079277.

Dankl, K. (2017) Design age: Towards a participatory transformation of images of ageing. *Design Studies*, Vol. 48, pp. 30–42. https://doi.org/10.1016/j.destud.2016.10.004.

Dave, S. (2010) High urban densities developing countries: A sustainable solution? *Built Environment*, Vol. 36/1, pp. 9–27. https://www.jstor.org/stable/23289981.

Debnath, A., Chin, H., Haque, M.M., and Yuen, B. (2014). A methodological framework for benchmarking smart transport cities. *Cities*, Vol. 37.

Docherty, I., Marsden, G., and Anable, J. (2018) The governance of smart mobility. *Transportation Research Part A: Policy and Practice*, Vol. 115, pp. 114–125. https://doi.org/10.1016/j.tra.2017.09.012.

Dong, L., Santi, P., Liu, Y., Zheng, S., and Ratti, C. (2022) The universality in urban commuting across and within cities. https://doi.org/10.48550/arXiv.2204.12865.

Duan, Y., Chen, X., Houthooft, R., Schulman, J., and Abbeel, P. (2016) Proceedings of the 33rd International Conference on Machine Learning, PMLR 48, pp. 1329–1338. https://doi.org/10.48550/arXiv.1604.06778.

European Commission (EC). (2011) Roadmap to a single European transport area—Towards a competitive and resource efficient transport system. EU white paper COM(2011) 144. Brussels: European Commission.

Eendebak, R., and World Health Organization. (2015) *World report on ageing and health*. World Health Organization.

Flipo, F., Dobré, M., and Michot, M. (2013) La face cachée du numérique. L'impact environnemental des nouvelles technologies. Le Kremlin Bicêtre: L'Échappée.

Fragkias, M., Lobo, J., Strumsky, D., and Seto, K.C. (2013) Does size matter? Scaling of CO_2 emissions and U.S. urban areas. *PloS One*, Vol. 8/6, pp. 1–8. https://doi.org/10.1371/journal.pone.0064727.

Geels, F.W. (2004). From sectoral systems of innovation to socio-technical systems: Insights about dynamics and change from sociology and institutional theory. *Research Policy*, Vol. 33/6, pp. 897–920. https://doi.org/10.1016/j.respol.2004.01.015.

Gehl, J., and Svarre, B. (2013) *How to study public life*. Washington, DC: Island Press.

Gehl, J. (2011) *Life between buildings: Using public space*, 6th ed. London: Island Press.

Gössling, S., Ceron, J.-P., Dubois, G., and Hall, C.M. (2009) Hypermobile travellers. In: Gössling, S., and Upham, P. (eds.), *Climate change and aviation*. London: Earthscan.

Green, B. (2020) *The smart enough city. Putting technology in its place to reclaim our urban future*. Cambridge: The MIT Press.

Hanson, S. (2010) Gender and mobility: New approaches for informing sustainability. *Gender, Place & Culture*, Vol. 17/1, pp. 5–23. https://doi.org/10.1080/09663690903498225.

Hargreaves, T., Longhurst, N., and Seyfang, G. (2012) Understanding sustainability innovations: Points of intersection between the multi-level perspective and social practice theory. 3S Working Paper 2012-03. Norwich: Science, Society and Sustainability Research Group.

Hensher, D.A., and Hietanen, S. (2023) Mobility as a feature (MaaF): Rethinking the focus of the second generation of mobility as a service (MaaS). *Transport Reviews*, Vol. 43/3, pp. 325–329. https://doi.org/10.1080/01441647.2022.2159122.

Hietanen, S. (2014) 'Mobility as a service'—The new transport model? *Eurotransport*, Vol. 12/2. ITS & Transport Management Supplement, pp. 2–4.

Hillier, B. (2009) Spatial sustainability in cities: Organic patterns and sustainable forms. Proceedings of the 7th International Space Syntax Symposium. Stockholm: KTH.

Ho, C.Q., et al. (2018) Potential uptake and willingness-to-pay for Mobility as a Service (MaaS): A stated choice study. *Transportation Research Part A*, Vol. 117, pp. 302–318. https://doi.org/10.1016/j.tra.2018.08.025.

Hoornweg, D., and Pope, K. (2017) Population predictions for the world's largest cities in the 21st century. *Environment and Urbanization*, Vol. 29/1, pp. 195–216. https://doi.org/10.1177/0956247816663557.

Huré, M. (2019) *Les mobilités partagées. Régulation politique et capitalisme urbain*. Paris: Editions de la Sorbonne.

IEA. (2023) *Global EV Outlook 2023*. Paris: IEA.

Insee. (2022) Causes de décès en 2017: Comparaisons régionales et départementales [Gov]. Institut National de La Statistique et Des Études Économiques (Insee). https://www.insee.fr/fr/statistiques/2012788.

ITDP. (2020) *What is TOD?* Institute for Transportation & Development Policy (ITDP). https://www.itdp.org/library/standards-and-guides/tod3-0/what-is-tod/.

ITF. (2021) Travel transitions: How transport planners and policy makers can respond to shifting mobility trends. ITF Research Reports. Paris: OECD Publishing.

Jabareen, Y.R. (2006) Sustainable urban forms: Their typologies, models, and concepts. *Journal of Planning Education and Research*, Vol. 26, pp. 38–52. https://doi.org/10.1177/0739456X05285119.

Jacobs, J. (1961) *The death and life of Great American cities*. New York (NY): Random House.

Kaack, L.H., Donti, P.L., Strubell, E., Kamiya, G., Creutzig, F., and Rolnick, D. (2022) Aligning artificial intelligence with climate change mitigation. *Nature Climate Change*, Vol. 12, pp. 518–527. https://doi.org/10.1038/s41558-022-01377-7

Kamargianni, M., Li, W., Matyas, M., and Schäfer, A. (2016) A critical review of new mobility services for urban transport. *Transport Research Arena, TRA2016*, Vol. 14, pp. 3294–3303.

Katzmarzyk, P.T., Friedenreich, C., Shiroma, E.J., and Lee, I.M. (2022) Physical inactivity and non-communicable disease burden in low-income, middle-income and high-income countries. *British Journal of Sports Medicine*, Vol. 56/2, pp. 101–106. https://doi.org/10.1136/bjsports-2020-103640.

Karfopoulos, E.L. and Hatziargyriou, N.D. (2012) A Multi-Agent System for Controlled Charging of a Large Population of Electric Vehicles. *IEEE Transactions on Power Systems*, Vol. 28/2, pp. 1196–1204. https://doi.org/10.1109/TPWRS.2012.2211624/

Khisty, C.J., and Zeitler, U. (2001) Is hypermobility a challenge for transport ethics and systemicity? *Systemic Practice and Action Research*, Vol. 14, pp. 597–613. https://doi.org/10.1023/A:1011925203641.

König, D., Eckhardt, J., Aapaoja, A., Sochor, J., and Karlsson, M. (2016) Deliverable 3: Business and operator models for MaaS. MAASiFiE project funded by CEDR.

Kropotkin, P. (1902) *Mutual aid: A factor of evolution*. London: Freedom Press.

Law, R. (1999) Beyond 'women and transport': Towards new geographies of gender and daily mobility. *Progress in Human Geography*, Vol. 23/4, pp. 567–588.

Le Breton, E. (2019). *Mobilité, la fin du rêve?* Paris: Éditions Apogée.

Lee, U., Kang, N., and Lee, I. (2020) Shared autonomous electric vehicle design and operations under uncertainties: A reliability-based design optimization approach. *Structural Multidisciplinary Optimization*, Vol. 61, pp. 1529–1545.

Lefèvre, B., and Mainguy, G. (2009) Urban transport energy consumption: Determinants and strategies for its reduction, S.A.P.I.EN.S. *Surveys and Perspectives Integrating Environment and Society*, Vol. 2/3, pp. 1–18.

Le Néchet, F. (2012) Urban spatial structure, daily mobility and energy consumption: a study of 34 European cities, Cybergeo: European Journal of Geography, Sistemas, Modelística, Geoestadísticas, 580. https://doi.org/10.4000/cybergeo.24966

Lesteven, G. et al. (2018). La transformation numérique des mobilités. Le nouveau monde de la mobilité, Presses des Ponts, pp. 141–146.

Li, H., Wan, Z., and He, H. (2019) Constrained EV charging scheduling based on safe deep reinforcement learning. *IEEE Transactions on Smart Grid*, Vol. 11/3, pp. 2427–2439. https://doi.org/10.1109/TSG.2019.2955437.

L'Institut Paris Region. (2019) *Cities change the world*. Paris: L'Institut Paris Region.

Liu, R., Dow, L., and Liu, E. (2011) A survey of PEV impacts on electric utilities. ISGT 2011, Anaheim, pp. 1–8. https://doi.org/10.1109/ISGT.2011.5759171.

Louf, R., and Barthelemy, M. (2014) Scaling: Lost in smog. *Environment and Planning B: Planning and Design*, Vol. 41, pp. 767–769. https://doi.org/10.1068/b4105c.

Marchetti, C. (1994) Anthropological invariants in travel behavior. *Technological Forecasting and Social Change*, Vol. 47/1.

Mnih, V., Kavukcuoglu, K., Silver, D. et al. (2015) Human-level control through deep reinforcement learning. *Nature*, Vol. 518, pp. 529–533. https://doi.org/10.1038/nature14236.

Montgomery, C. (2013). *Happy city: Transforming our lives through urban design*. New York: Farrar, Straus and Giroux.

Mouly-Aigrot, B., Fouco, L., Leurent, F., and Lesteven, G. (2016) La transformation numérique nouvel eldorado pour les acteurs des transports?

Mulley, C. (2017) Mobility as a Services (MaaS)—Does it have critical mass? *Transport Reviews*, Vol. 37/3, pp. 247–251. https://doi.org/10.1080/01441647.2017.1280932.

National Academies of Sciences, Engineering, and Medicine. (2016) *Between public and private mobility: Examining the rise of technology-enabled transportation services*. Washington, DC: The National Academies Press. https://doi.org/10.17226/21875.

Newman, P., and Kenworthy, J. (1989) *Cities and automobile dependence: An international sourcebook*. Aldershot: Gower.

Nussbaum, M. (2003) Capabilities as fundamental entitlements: Sen and Social Justice. *Feminist Economic*, Vol. 9/2–3, pp. 33–59. https://doi.org/10.1080/1354570022000077926.

Oliveira, V. (2016) *Urban morphology: An introduction to the study of the physical form of cities*. The Urban Book Series. Basel: Springer International Publishing.

Oliveira, E.A., Andrade, J.S., and Makse, H.A. (2014) Large cities are less green. *Scientific Reports*, Vol. 4, pp. 1–12. https://doi.org/10.1038/srep04235.

Paap, J., and Katz, R. (2004). Anticipating disruptive innovation. *Research-Technology Management*, Vol. 47/5, pp. 13–22.

Picon, A. (2015) *Smart cities: A spatialised intelligence*. Chichester: Wiley.

Pont, M.B., and Haupt, P. (2009) Space, density and urban form. Doctoral Thesis, Delft: Technical University Delft.

Randall, T. (2016) Here's how electric cars will cause the next oil crisis. A shift is under way that will lead to widespread adoption of EVs in the next decade. Bloomberg New Energy Finance 25.

Reyes Madrigal, L.M., Nicolaï, I., and Puchinger, J. (2023) Pedestrian mobility in Mobility as a Service (MaaS): Sustainable value potential and policy implications in the Paris region case. *European Transport Research Review*, 15/13. https://doi.org/10.1186/s12544-023-00585-2.

Rifkis, J. (2011) *The third industrial revolution*. London: Palgrave Macmillan.

Rittel, H.W.J., and Webber, M.M. (1973) Dilemmas in a general theory of planning. *Policy Sciences*, Vol. 4, pp. 155–169.

Robomobile Life Workshop. (2020) Prospective Atlas of the Robomobile Planet. The robomobile life. Prospective workshop. Paris: La Vie Robomobile.

Rothenberg, S. (2007) Sustainability through servicizing. *MIT Sloan Management Review*, Vol. 48/2.

Sarasini, S., Sochor, J., and Arby, H. (2017) What characterises a sustainable MaaS business model? ICoMaaS 2017 Proceedings, 2017.

Sen, A. (1979) *Equality of what? The Tanner lecture on human values*. Stanford University, 22 May 1979.

Singh, V.P., Kishor, N., and Samuel, P. (2017) Distributed multi-agent system-based load frequency control for multi-area power system in smart grid. *IEEE Transactions on Industrial Electronics,* Vol. 64/6, pp. 5151–5160.

Smith, G. (2020) Making mobility-as-a-service: Towards governance principles and pathways. Doctoral Dissertation. Chalmers University of Technology.

Sochor, J., Arby, H., Karlsson, I.C.M., and Sarasini, S. (2018) A topological approach to mobility as a service: A proposed tool for understanding requirements and effects, and for aiding the integration of societal goals. *Research in Transportation Business & Management*, Vol. 27, pp. 3–14. https://doi.org/10.1016/j.rtbm.2018.12.003.

Stead, D. (2016) Identifying key research themes for sustainable urban mobility. *International Journal of Sustainable Transportation*, 8318(February), Vol. 10/1, pp. 1–8.

Streeting, M., Chen, H., and Edgar, E. (2017) Mobility as a Service: The Next Transport Disruption. Special Report. LEK Consulting. International Association of Public Transport (UITP) Australia & New Zealand and Tourism & Transport Forum (TTF) Australia.

Townsend, A. (2014) *Re-programming mobility: The digital transformation of transportation in the United States*. New York: The NYU Wagner Rudin Center.

UN-Habitat. (2022) Envisaging the future of cities. World Cities Report 2022. Nairobi: UN-Habitat.

UNEP. (2008) Urban density and transport-related energy consumption. UNEP/GRID-Arendal Maps and Graphics Library.

Wang, D., Coignard, J., Zeng, T., Zhang, C., and Saxena, S. (2016) Quantifying electric vehicle battery degradation from driving vs. vehicle-to-grid services. *Journal of Power Sources*, Vol. 332, pp. 193–203.

Woodcock, J., Edwards, P., Tonne, C., Armstrong, B.G., Ashiru, O., Banister, D., Beevers, S., Chalabi, Z., Chowdhury, Z., Cohen, A., Franco, O.H., Haines, A., Hickman, R., Lindsay, G., Mittal, I., Mohan, D., Tiwari, G., Woodward, A., and Roberts, I. (2009) Public health benefits of strategies to reduce greenhouse-gas emissions: Urban land transport. *The Lancet*, Vol. 374/9705, pp. 1930–1943. https://doi.org/10.1016/S0140-6736(09)61714-1.

Xie, X.-F., Smith, S., and Barlow, G. (2012) Schedule-driven coordination for real-time traffic network control. Proceedings of the International Conference on Automated Planning and Scheduling.

CHAPTER 3

Approaches for Sustainable Urban Mobility Futures

Abstract Aside from the trends, there are some dominant—and promising—ways of addressing today's problems whilst preparing for the developments of tomorrow. In this chapter, we present and discuss three types of approaches and some of their methods. First, we look at urban governance. Whilst not at the focus of this book, it is the foundation for many of the other approaches and a key element of urban mobility system transitions. Next, we outline people-centred design approaches, including some examples of how that can be achieved in future settings. Lastly, we look at data-driven design and decision-making with in-depth discussions of mobility system modelling and charging infrastructure management.

Keywords Methodology · Urban governance · Design · Urbanites · Decision-making · Simulation · Modelling

In the same way as we can identify different categories of trends and uncertainties, there are some dominant—and promising—ways of addressing today's problem whilst preparing for the developments of tomorrow. In this chapter, we present and discuss three types of approaches and some of their methods. Each of the approaches and methods responds to one or several of the challenges and opportunities discussed in the previous chapter. Whilst they are discussed separately

T. Gall et al., *Sustainable Urban Mobility Futures*, Sustainable Urban Futures, https://doi.org/10.1007/978-3-031-45795-1_3

here, they are often used in parallel or consecutively. In this section, we present them through specific examples of transitions from Paris, whilst also looking at some challenging cases to emphasise the risks. In the following chapter, we build on the trends and uncertainties from the first chapter and show possible pathways how the presented approaches can—in an integrated and holistic approach—support the design and planning of sustainable and people-centred urban mobility systems.

Possible ways to reformulate the approaches towards sustainable transport and people-centred urban mobility systems on a European scale started to consolidate in 2011, when the European Commission published a white paper in which two priorities for the future of transportation were established: competitiveness and efficiency. The proposed strategies to achieve these objectives included optimising multimodal logistics chains, developing information systems, and providing incentives to invest in these markets. The European Commission also emphasised the key role of innovation and collaborative stakeholder participation towards sustainable transportation organisation (EC 2011).

Strategies to ensure access to mobility for all have also evolved over time. The aim is no longer to build urban highways to provide 'access' to territories, but rather to go beyond individual motorisation. The pursuit for accessibility in urban mobility is encouraged by several factors: the climate crisis, the economic crisis, metropolisation, the growing size of urban cores, and the institutional, spatial, and organisational challenges that all of these imply (Salet et al. 2003).

The interaction of all elements in urban mobility systems is closely linked to social (housing, leisure) and economic (work, production, etc.) dynamics that coexist at various scales of territories and powers. These dynamics give mobility systems their own characteristics and diverse functions (Watson 2012). The analysis of mobility systems has evolved: They are no longer approached through a dichotomous analysis of car vs. public transport, but under a broad spectrum that considers various new and old mobility services. The complexity of mobility choices could be explained, on the one hand, by the multiplication of service offers linked to an evolution of demand. On the other hand, by technological innovations that have enabled the development of diverse transport solutions.

The new interactions between multiple mobility services in urban transport systems now seem to depend on political choices (or the absence of choices, e.g., the arrival of electric scooters or its banning in Paris) and on developments in local and global economic contexts (e.g., gentrification, metropolisation). The multiplication of mobility choices amplifies the existing challenges of competition for road use and competition between modes of transport.

3.1 Enabling Urban Governance

The concept of governance is defined as 'a process of actor, social, and institutional coordination to achieve collective objectives' (Courmont and Le Galès 2019). The governance concept outlines the collaboration between actors from many sectors and hierarchies working together to build a common project. A well-structured governance facilitates the collective consensus to build strong policies (Docherty et al. 2018).

Public–private governance dynamics have re-emerged as market opportunities at the end of the twentieth and the beginning of the twenty-first century. They are considered as strategies aimed at improving the efficiency of public services and reducing public budget expenditures in austerity policies (Courmont and Le Galès 2019; Pangbourne et al. 2018).

The innovation in the actor dimension involves new expectations, processes, and dynamics to govern mobility (Lascoumes and Le Galès 2005). Governance dynamics seem to reinvent themselves. New interactions, rich in public–private partnerships and other types of cooperation, aim to better respond to material, environmental, and economic needs for all stakeholders in the urban ecosystems. The encouragement of partnerships and certain contractual processes has been reinforced by a dominant European political will in terms of opening competition amongst actors within and outside the current mobility ecosystem (EC 2011).

The American crisis of the thirties can be seen as one of the major turning points in the evolution of governance processes in cities, with the inclusion of new partnerships that took into account the interests of public–private partners, shareholders, and society, as contextualised by Charreaux (2004). New processes have been consolidated with regulatory objectives to frame the actions of actors and establish new 'managerial rules of the game' (Charreaux 2004). In addition, in France, the post-decentralisation years around 1983 could be seen as a key moment for the construction of new governance dynamics across institutional scales.

Current governance processes in urban mobility systems seem to become more complex to achieve collaborative decision-making towards resource optimisation for all stakeholders: individual or collective actors;

public, private, or semi-private (Piatoni 2010). The design of governance can be understood in this work as the configuration in systems of actors and institutions for negotiation and decision-making. These configurations are in continuous evolution, meaning that they are dynamic processes that would allow for better results at the organisational and economic levels.

The integration of urban public transport in a diverse system of actors, start-ups, multinationals, and in a diversified cost system, makes the organisation of public transport systems an issue that requires various mechanisms to improve collaboration amongst all the involved actors towards common objectives. The European regulatory framework for transport and mobility system organisation gives the responsibility to the competent authority to offer public transport services through market tools translated into contracts with transport operators. The following sections zoom in on three dimensions of urban governance: The regulation of public transport, urban data, and urban governance interventions.

Regulation of public transport

The dynamics of social complexity and mobility in urban environments share similar challenges globally: actor reconfigurations, financial crises, and the climate crisis. The role of institutions in managing these new complex urban realities (e.g., multiple actors, budgetary issues, environmental challenges) is important. It ensures continuity during crises and prevents them from compromising the evolution and development path of cities and their mobility systems.

The analysis of the stakeholder's ecosystem is necessary to understand the context, choices, and socio-economic interests that have an impact on the organisation, financing, and regulation of mobility systems). Legislative supports provide the legal framework and evolve to take into account technological innovations that are incorporated into mobility systems (Lajas and Macário 2020). However, the construction of regulatory frameworks to oversee mobility innovations is a current issue, as the processes for constructing these frameworks seem to be outpaced by the speed of innovation (Smith 2020). Financing public transport is another challenge related to the institutional framework of public transport authorities, given that organisations in charge of organising mobility are most often also responsible for ensuring their economic balance. The origin of resources to finance transportation is a topic that often appears on political agendas. This can vary between heavily subsidised and free public transport, defunding of public transport due to 'economic underperformance', or a prioritisation of resources based on the visibility that beneficiary services give to elected officials.

In Europe, this organisation of public transportation systems is usually the responsibility of local or regional public authorities. The organisations in charge of transport planning are the mobility authorities (Public Transport Authorities (PTA), regulated in France by the mobility orientation law, LOM, from December 2019). These authorities are responsible for organising public transport services, regulating them as well as the interactions between various operators. They define and operate service coverage and quality objectives according to the territorial economic context (Roy and Yvrande-Billon 2007; Smith 2020). The operating companies that offer their transport services (e.g., buses, trams, trains, metro) in the urban public transport network under the regulation of the PTA are usually also responsible for maintaining vehicles, ticketing, network maintenance, and transport service operations. Urban public transport services are frequently provided by companies in a monopolistic or oligopolistic situation, as is the case, for example, in Paris, Helsinki, and Vienna (Reyes Madrigal 2021). Public authorities also have the role of regulating competition between modes. Two examples of forms of public transport service management are: direct operation, where the public authority plays the role of operator/manager; public service delegation (DSP), either to a private company or to a mixed company (cf. SEM or other type of syndicate) (Roy and Yvrande-Billon 2007).

Regulation of urban data

A second interesting field is that of urban data governance. The management of data in urban environments has taken centre stage across sectors, mobility being one of the dominant ones due to the number of actors involved as well as the potential of use of the data. The policies and frameworks that underpin this are shaped across levels, including European Union directives, and national regulations. The two dominant movements are the development of data standards and the rise of open data initiatives. Both can origin from the public and private side. OpenStreetMap is a leading example for crowd-sourced and open spatial data now widely used for private, commercial, and public projects. The General Transit Feed Specification (GTFS), the basis for an interoperable standard for public transport schedules, originated from Google but has spread to all sectors. On the other hand, public regulations, such as the French mobility law LOM, increasingly push for open data and make the provision of data mandatory for actors in the mobility ecosystem. In a world where increasingly more decisions are built on data, the availability and quality of data is crucial. Within this, lots of work is ongoing regarding ensuring data quality, clarifying ownership and rights, standardisation, as well as novel data collection methods such as crowdsourcing or citizen science. In essence, urban data governance in the mobility sector refers to a blend of policies, data practices, and public engagement.

Urban governance interventions

Lastly, an increasing number of governance interventions targets the sustainability transition of urban mobility systems. This can take various shapes. For example, ambitions by various cities—following the example of Paris—advocate for 15-minute cities where everything necessary for daily life is in close proximity. A much stricter instance is the concept of Low-Emission Zones (LEZ), wherein certain areas restrict the entry of vehicles failing to meet emission standards. This policy can be a part of Sustainable Urban Mobility Plans (SUMP), another policy instrument gaining momentum, and reflect a broader push towards sustainable urban mobility policy. Additionally, policies involving taxation, via tolls, parking fees, or other financial incentives or costs, are part of the urban policies field. Many of them aim to transform the behaviours of people, either via financial or information-based measures. The book 'Transport for Humans' by Dyson and Sutherland (2021) provides a holistic perspective on this approach, highlighting its significant potentials.

This completes the first section on approaches for sustainability urban mobility futures. The next section continues from the behavioural lens focusing on individuals and users and presents a range of concepts and methods for people-centred design.

3.2 People-Centred Design—Design for Future Urbanites

In this section, we advocate that future-oriented approaches initially relied on resource optimisation, then moved to tech-oriented reflections, and became more recently user/human-centred, to people-centred. Integrating a people-centred perspective in urban mobility system design and the creation of scenarios is seen as a key activity for transitions pathways. The advantages are twofold. It permits to imagine and enable sustainable mobility behaviours for different social groups and prepare for different mobility behaviours associated with future living conditions. An important dimension is the individual traveller's experience which defines choices of more or less sustainable modes or evolving behaviours (see Al Maghraoui et al. 2019, for example, on traveller experience). The notion of *mobility justice* presented in SDG 11.2 or, at a national level, in mobility laws (see for instance the LOM law in France) can be addressed partially through these methods.

3.2.1 Variants of People-Centred Design

A variety of design methodologies interacting, in some way, with humans can be grouped through people-centred design. The term is constituted of the meaning of the word—design with people, or human characteristics at the core, complemented with the definition of being foremost a non-linear, impact-driven approach and process which solves problems by involving people in the process and creating outcomes which are responding to the human's expectations (IDEO 2015). Multiple people-centred design approaches have been developed, such as user-centred, people-centric, inclusive or participatory design. They are applied across various design disciplines such as product and service design, architecture, or urban planning. Yet, they vary significantly in their characteristics and consider the human to varying degrees and with different methods.

This section provides an overview over existing approaches from the design and planning field and an initial classification. The extended definition and description of each type can be found after the classification. The focus lies on approaches directly linked to a varying level of interaction or influence on people. We group methods with similar approaches and use to the most common name. Existing classifications and frameworks are the foundation for the following (Sanders and Stappers 2008, 2014; Buur and Matthews 2008).

Most notably, Sanders and Stappers (2008) developed a matrix divided into 'led by design' vs. 'led by research' on one axis and 'user as subject' vs. 'user as partner' on the other. User-centred design approaches are situated on the passive, research-led side, encompassing usability testing, human factors and ergonomics, applied ethnography, lead-user innovation, and contextual inquiry. On the other hand, participatory design research falls onto the active side, including generative design research and 'Scandinavian'. The latter refers to one of the first participatory practices centred around 'deep commitments to democracy and democratisation; discussions of values in design and imagined futures; and how conflict and contradictions are regarded as resources in design' (Gregory 2003).

One possible dimension of classification addresses the level of participation in the design process as proposed by Sanders and Stappers (2008). This is relevant for two reasons. First, the definition of people-centred design focuses on the integration of humans in the process as a core element. Second, the level of participation of humans may relate to the impacts of the product or service and is, therefore, central for the purpose of this paper. Hence, we categorise the approaches ranging from no participation to co-creation, and finally decentralised design (user designs independently, described in detail below).

Figure 3.1 portrays the different people-centred design approaches. The x-axis portrays a qualitative assessment of the different approaches' range of impact on people, ranging from the user human on the left, to people/humankind on the right. On the y-axis, it shows the range of typical levels of participation, ranging from no/low participation to high levels of participation.

Whilst the arrangement in Fig. 3.1 is not irrefutable, it can lead to three observations: (1) More active people-centred design processes

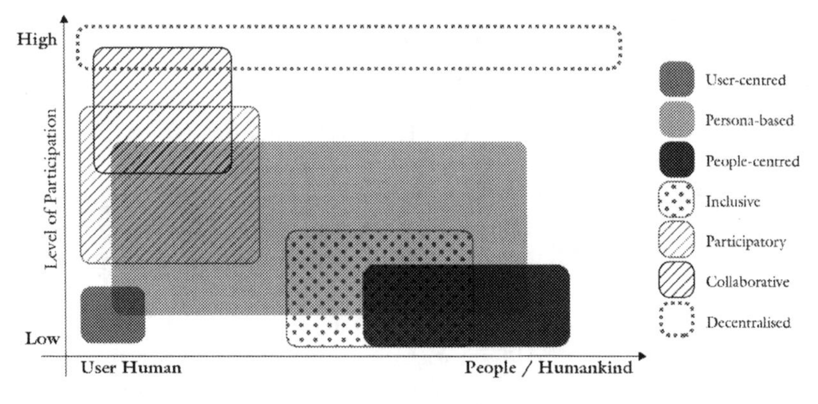

Fig. 3.1 People-centred design approaches (Adapted from Gall et al. [2021])

are likely to focus on impacts on user humans; (2) Approaches more focused on people are less likely participative; (3) None considers fully the impacts on humankind, even less so from the more participatory approaches. Excluded from this is decentralised design, which is most participative but does not necessarily consider anyone but the designing human itself. Yet, if everyone has the same capability of designing for themselves, it could lead to a democratic representation of each human, hence humankind at large. However, as it is not a widespread method, it cannot be fully considered. To expand on the characteristics, the following section describes and compares the methods.

User-centred design

User-centred design can be defined as 'an iterative design process in which designers focus on the users and their needs in each phase of the design process' in which 'design teams involve users throughout the design process via a variety of research and design techniques, to create highly usable and accessible products for them' (Interaction Design 2020). The approach is often linked to user interface design and other sub-fields which intend to be responsive and/or predictive, e.g., through the use of market analyses or weak signals. The approach can be supplemented by concepts such as future users (Gregory 2003), lead user/early adopters/ lighthouse customer (Buur and Matthews 2008), or methods such as customer journeys. Empathic or compassionate design add an additional dimension (Seshadri et al. 2019).

People-centric design

People-centric design is a variation of user-centred design with a stronger focus on the people and the public perspective. It originates from the field of urban design, spearheaded by Danish architect and urbanist Jan Gehl (2011) in the 1960s. He studied public spaces in Italy through a multi-disciplinary lens, leading to a renaissance of concepts such as human-scale and human-friendly cities in times of car-dominated urban transformations. His work constituted the European pendant to Jane Jacobs (1961) pioneering work to re-focus cities on people through a list of recommendations which remain valid until today. Since then, conceptual subsidiaries arose, e.g., child-friendly design. In a broader sense, people-centric design can be grouped with design anthropology (Buur and Matthews 2008).

Inclusive design

Inclusive design highlights that each 'design decision has the potential to include or exclude customers' and focuses on 'the contribution that understanding user diversity makes to informing these decisions' (University of Cambridge 2023). Whilst often used to design for people with disabilities, this would be more accurately accessible design. Further, inclusive does not equal inclusionary design. The former allows everyone to participate, whereas the latter refers to prioritising the diversification and consideration of unique groups. Compared to the previous approaches, inclusive design shares many similarities but adds the importance for individuality in observing and including people's needs in design processes to reach good outcomes for all groups.

Participatory design

Participatory design refers to design processes in which the user directly participates. This can manifest in a variety of ways, from short opinion surveys at the start of the process to an ongoing consultation. Other concepts with a similar meaning are citizen/user/stakeholder engagement, amongst others (Bertolini 2020). Whilst participatory design is centred on the active involvement of people, the design process remains separate and only informed by the participation process. Therefore, participation can sometimes become a mere box to tick or mandatory element without integration of the results in subsequent steps.

Collaborative design

The next level of active involvement is co-design, co-creation, or co-production. The prefix 'co' refers to collaborative, or the simultaneous, open, and horizontal collaboration between designers/experts and users/citizens. Extended conceptualisations thereof include the co-production as part of co-creation, e.g., through storytelling (Gall and Haxhija 2020), or practices of mediation, negotiation, and consensus finding (Watson 2002, 2003). Whilst it goes under various names, co-design and co-creation found widespread use and are often jointly developed with initiatives towards open science, citizen science, or citizen observatories. Whilst the user plays a significant role, the designer remains a central figure as facilitator, mediator, or translator (of knowledge) (Sanders and Stappers 2008; Verloo 2019).

Decentralised design

Lastly, and less formally defined, an approach of decentralised design has been added in response to current trends such as decentralised or newly re-localised production as well as the concept of prosumers (acting as both producers and consumers), enabled by new technologies and values (e.g., Kropotkin 1902; Rifkin 2011). The proposed addition describes a process in which no designer or external expert is actively involved in. The role of a facilitator is rendered substitutable, for example, due to improved interaction with machines through simplified processes. Whilst this does not apply in many cases yet, it might in the future. A common example is the potential of decentralised design and production enabled through 3D-printing (Urry 2016). However, decentralised design describes a field broader than the existing maker movement and other initiatives that evolved around 3D printing. When imagining design approaches of the future, the concept of post-automation can add another layer, describing the period where most essential functions would be automatised and people at large would have access to design and production facilities, an abundance of time and interest or even need to create and produce artifacts.

Another analogy can be drawn to approaches such as bottom-up, community-led, or grassroots design (Seyfang and Smith 2007). Compared to approaches by larger organisations or institutions, they start like decentralised design on a very small local scale. However, they intend to scale up. This can be seen as a variation or potential consequence of decentralised design, which in itself just focuses on the design process of an individual for an individual or a very small group (such as the family).

Persona-based design

Another approach, building on user-centred design, is persona-based design. Initially developed by Cooper (1999) as user personas for software development, teams within Microsoft extended the application and set out foundations which find their application until today (Pruitt and Grudin 2003). User personas have been extended in various ways, for examples, towards future or anti personas (Fergnani 2019; Fergnani and Jackson 2019; Fuglerud et al. 2020; Miaskiewicz and Kozar 2011; Salminen et al. 2020). They can be developed in various ways depending on available information, resources, and requirements profiles. These can be through focus group, survey-based (Schäfer et al. 2019) or big-data-based (Stevenson and Mattson 2019). Whilst personas bear potentials, a challenge is that they can never represent the diverse reality. However, they can contribute to more accentuated representation of users. Additionally, they can ensure that statistically underrepresented groups such as people with disabilities are considered through a disproportional portrayal. Whilst this approach is traditionally either a variation of user-centred or people-centric design, it can be seen as a more inclusive and participatory approach depending on the chosen process (Vallet et al. 2020). If personas are co-created but only used by designers, it can be an inclusive process. If the created personas inform a collaborative design process, it can merge the strengths of various tools.

3.2.2 Combining Persona-Based Design with Scenario Planning

In scenario planning, future scenarios are expected to spark engagement and empathy from the users. Introducing characters who come to life in the scenarios is one method to do so, yet rarely discussed and implemented in research and practice (Fergnani 2019; Vallet et al. 2020). To address this gap, Fergnani (2019) introduced the notion of 'future persona'. The 'future persona' is scenario-specific, i.e., attached to one alternative scenario. For the author, it is easier for the reader to mentally connect to one single character per scenario. A second way is to illustrate a normative transition pathway through the evolutive description of a set of representative personas (Elioth 2017). This was developed in a prospective exercise for the Île-de-France region in 2016 to illustrate a carbon neutrality pathway: to fit the 2 °C target, 50% less CO2e emissions are expected in 2030 and 80% less in 2050 (compared to 2004 levels). A narrative involving 13 characters was created (Fig. 3.2). In the storyline, the personas change profiles (and associated values) over time between 2016 and 2050. Some of them meet, share their life whilst others pass away. Histograms of estimated individual C02e emissions are provided to account for a change: from a mean 10 tCO2e/person in 2016 to 1.8 tCO2e/person in 2050. A third view is to consider the variations of a personas across future scenarios showing how, with an identical set of primary beliefs, their activities differ in the various scenarios (Vallet et al. 2020).

Personas	Stéphanie	Monique	Jacques	Leila	Julien	Eric	Nadia	Olga	Camille	Emilie	Manuel	Thierry	Adnan
Age	41	65	68	28	32	53	53	77	22	34	39	50	22
Position	Care assistant	Retired		IT project manager		Journalist	Asset manager	Retired	Student	Teacher	Civil servant	Job seeker	Syrian refugee
Children	1	0		0		2		0	0	1		0	0
Status	Modest	High		Average		Very high		Average	Modest	Average		Modest	Very modest
App. size	44 sqm	90 sqm		45 sqm		256 sqm		60 sqm	16 sqm	56 sqm		36 sqm	15 sqm
Arrondis.	13th	4th		9th		7th		16th	19th	2th		14th	18th
Housing	S	-		R		-		-	SH	-		S	H-T
Short trips	Car	Car		Petrol scooter		Car	Taxi	Bike	Public transport	Public transport (PT)		Walk/PT	Walk
Long trips	Train	Car	Plane	Train	Plane	Car	Plane	Plane	Carsharing train	Car	Plane	Car pooling	HiH Carsharing
Nutrition	Carni-vore	Omnivore		Carnivore		Carnivore		Carni-vore	Flexiteri-an	Flexiterian		Carni-vore	Omni-vore
Cluster	C6	C5		C1	C2	C8	C6	C5	C1/C8	C7	C2	C8	C8

SH: Social housing; R: Rent; S: Shared; H: Hostel; T: Tent; PT: Public transport; HiH: Hitchhiking

Fig. 3.2 Persona development for Île-de-France in 2016 (Vallet et al. 2022, based on Elioth [2017])

To introduce in a systematic and qualitative way individual perspectives projected on future mobility scenarios, we present the Scenario Personanarrative method (Vallet et al. 2020). Box 3.1 describes different ways and elements to keep in mind for generating future scenarios whilst Box 3.2 presents the method itself.

Numerous studies have focused on the future of mobility in the medium term (2030–2040) or in the longer term (2050 or 2100). For the example here, we restrict the field of study to three technological and behavioural revolutions that are well represented in the literature (in addition to the simple mobility of people): electric mobility or electromobility, autonomous mobility and shared mobility (see section on *Technological trends*). In addition, we retain the prospective works that propose two to four scenarios of societal evolution. After selecting a sample of eleven representative studies published between 2011 and 2016, we extracted the objectives, main variables, and envisaged impacts. A qualitative analysis followed by a clustering session revealed common features in the construction of the scenarios. We identified three archetypal scenarios, which we call: (1) Continuation of the current trend, (2) Acceleration and high-tech society, and (3) Deceleration and societal change. The three archetypal scenarios chosen are finally constructed on the basis of the Rohr et al. (2016) study for the UK.

Creation of profiles of future urbanites

We use three theories from various disciplines, originally developed for fields other than mobility. Each of them sheds light on the criteria for changing people's behaviour. These are Social Practice Theory (Reckwitz 2002), the Sinus Milieu approach (Bertram and Berthold 2012), and the Behaviour Change Model (Fogg 2009). Social Practice Theory focuses on the determinants of everyday routines. The Sinus Milieu approach focuses on the values, beliefs, and views of individuals. Finally, the Behaviour Change Model highlights the conditions necessary for behaviour change. The two main criteria that condition behaviour change are (1) the ability to adopt a particular behaviour, or 'skills' in a broad sense (e.g., education or financial means) and (2) the willingness, or motivation (internal or external) to adopt this type of behaviour. This leads to four profiles, combining the criteria of competences (limited or important resources) and motivation (in this case, conservatism or trendsetting). The mottos of the various profiles are expressed as follows: (1) Securing the state, (2) Paving the way, (3) Live for the moment, and (4) Leading in class.

Creation of short stories where future urbanites evolve in archetypal scenarios

The so-called persona model, or fictional character representative of a class of user (Cooper 1999) is frequently used in people-centred design. This model has the advantage of illustrating and concretising the profiles of future urbanites described above.

The systematic creation of mobile life stories 'A day in the life of X' is achieved by crossing the personas corresponding to the different profiles with the three archetypal societal scenarios. The framework for the creation of these stories is a two-hour participatory workshop, which brought together twelve practitioners in innovation as well as research and development in the field of energy and mobility. During the workshop, three of the four personas were addressed by three groups, consisting of four people each (including a facilitator). For the fourth persona, one of the groups did the same work a week later, under similar conditions. Details on the future persona creation are given in Box 3.2.

The stories that have been developed are illustrations of the effects of technological developments on people's lives, understandable by designers and policymakers alike. The potential effects on different social groups are made more tangible.

Box 3.1 Scenario generation [method]

Designing solutions today must respond to the uncertain needs and contexts of tomorrow. To integrate the future dimension in decision-making and design today, various more or less structured foresight methods exist. Instead of trying to predict—meaning assuming something will take place—foresight tries to integrate the unknown future varieties in the process. One of the most common approaches is using scenarios. They permit working strategically with multiple possible futures or plausible alternatives. The term is used widely with diverse meanings. However, building on an extensive review, Spaniol and Rowland define the key characteristics of scenarios as having 'a temporal property rooted in the future', considering 'external forces', and remaining 'possible and plausible whilst taking the proper form of a story or narrative description'. Finally, scenarios 'exist in sets that are systematically prepared to coexist as meaningful alternatives to one another' (Spaniol and Rowland 2018, p. 1). These meaningful alternatives are also called exploratory scenarios and shall assist designers and decision-makers today in preparing for the future (Börjeson et al. 2006).

Scenarios can be created with various approaches, depending on the needs and resources. One of the most common ones is the 2*2 approach. For this, uncertainties are collected and rated by their potential impact and likelihood. Two of the most likely and impactful ones are finally chosen. For the urban mobility context, that might be if mobility services will be shared or not, and if future solutions will be primarily technological or focusing on active modes. When putting both of them on x and y axes, we get four quadrants, each with a particular combination: (1) Low sharing, active modes, (2) High sharing, technology, (3) High sharing, active mobility, and (4) High sharing, technology. This builds afterwards the basis to detail each of the scenarios. In this process, other trends and uncertainties can be integrated by aligning them with the critical uncertainties. For example, the scenario 3 'High sharing, active mobility' would be defined by few personally owned vehicles but various sharing opportunities and lots of walking and cycling. Thus, a focus could be on walking and cycling on shared bikes to public transport stations or mobility hubs, various business models to co-own vehicles within a neighbourhood, e.g., a cargo bike that can be used to get groceries, and few cars shared for special purposes. Using the narrative, a distinct name can be given, the scenario can be localised in a specific setting, e.g., a city district, visualised via sketches, videos, future habitant interviews, amongst others, and described with quantified indicators, such as assumed modal shares (how many people use the bike, car, public transport) or average CO2e indicators.

Another, more resource-effective, approach is the reuse of existing scenarios. These can be more generic—e.g., national urban development or mobility scenarios—that are adapted to the local context or archetypical scenarios. The latter are usually derived by studying existing scenario sets and grouping them by common characteristics. In the mobility context, Miskolczi et al. (2021) have done so for scenarios until 2030, resulting in four scenarios: 'Good old transport', keeping things as they are today, 'Mine is yours', prioritising servitisation and sharing practices, 'At an easy pace', assuming a continuing transition towards more active solutions, and 'Tech eager mobility', putting technology first.

Important to keep in mind for scenarios are three things:

1. Only a set of plausible scenarios permits a true consideration of future uncertainty today,
2. Scenarios cannot be 'correct' but need to be effective in integrating uncertainty in the users' imagination and way of thinking, and
3. Despite the resource-intensive character, simple ways to integrate uncertainties exist and are always better than using an individual future prediction which is more likely to be false than true.

Variations of this approach were on three occasions: (1) A future mobility 2030 workshop of the Anthropolis chair with twelve research and development and mobility innovation practitioners, (2) The future mobility 2030–2050 workshop organised for the project 'Les routes du futur du Grand Paris - Forum Métropolitain du Grand Paris', and (3) The future mobility 2030 workshop with 20 students from the BEST European summer school (Table 3.1). So far, we have explored the approach with around fifty participants from a wide range of backgrounds, facilitated the creation of 13 personas and 42 short stories of future mobility in 2030 and 2050.

The common basis for the three workshops is the values and resources dimensions, which give substance to the personas. In the first two workshops, we imposed the age-profile combination, and the construction of the personas was purely qualitative and intuitive, based on fictitious elements. In workshop 3, the students were asked to draw profile dimensions at random to add a more playful dimension to the activity. A variant was observed in workshop 2: a semi-qualitative construction, as the three personas (and their children) were inspired by 20 short interviews with real people met in the vicinity of the northern Paris ring road, the study area for the Greater Paris Forum project.

To save resources, we chose to re-use existing scenarios from the literature on mobility. For workshop 2, the three generic scenarios identified in the first workshop were adapted to the specific characteristics of the Île-de-France region during preparatory work by a small facilitation team (architects, urban planners, mobility experts). To limit the cognitive load on participants during the workshop, a facilitator was tasked with giving a pitch of the scenarios at the start of each creative session.

The participants of the pilot workshop wrote mobility stories of a few paragraphs, which were shared in the group. Two complementary formats were then explored: rewriting, consolidation of the stories, and illustrating each story as a comic strip by a graphic artist, followed by presentation of the results of the Greater Paris Forum project including theatrical staging in a few minutes by summer school students (Bornet and Brangier 2013).

Table 3.1 Examples of workshops combining persona generation in future scenarios

Workshop	*1: Anthropolis workshop*	*2-Greater Paris workshop*	*3-International student workshop*
Participants	12 research and development practitioners (innovation and mobility)	12 professionals (architects, designers, mobility experts) 3 interns	20 European students
Duration	2 hours	2.5 hours	2.5 hours
Number of groups	4	3	6
Horizon	2030	2030 and 2050	2030
Scenario names	More of the same Acceleration Deceleration	Business as usual 2030/2050 Hypermobility 2030/2050 Hypomobility 2030/2050	(More of the same) Acceleration Deceleration
Personas-Fixed inputs	Age in 2030 Profile	Name-Couple of personas parent–child Age in 2020 Profile Location of residence (Northern area of Paris)	Couple of personas Age in 2030
Personas—Variables defined in the groups	Name, professional position	Professional position	Name Profile (random) Physical trait (random)
Narrative outcomes	12 stories (4*3) 10–15 lines/scenario/persona Orally exposed	18 stories (3*6) 5–10 lines/scenario/persona + audio recording Graphic post-treatment: 3 cartoon strips per persona 2030–2050	12 stories (6*2) 5–10 lines/scenario/persona Theatre presentation

Box 3.2 Scenario Personanarrative [method]

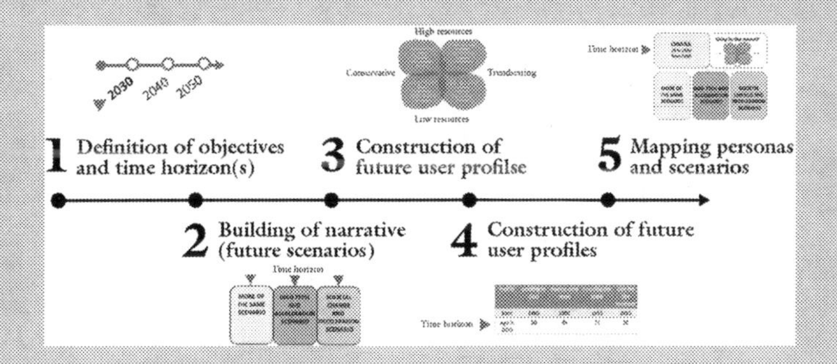

To exemplify the Personanarrative approach (Vallet et al. 2020), we present one of four personas that have been developed during one of the workshops. Each persona belongs to one of the four profile quadrants combining resource and values characteristics. The age and gender of each persona were chosen freely by each group charge. The four personas are Pierre-Antoine (M), from a higher socio-economic category, and conservative who will be 50 years in 2030. Zoe (F) is from a similar socio-economic category, considered a trendsetter, and 35 years old in 2030. Pascal has limited resources, is rather conservative, and will be 75 years old in 2030. Finally, Chiara (F), also with limited resources but a trendsetter, will be 20 years old in 2030. The following is an example of the baseline scenario for her, and how Chiara's mobile life could develop in the baseline scenario compared to one of the other scenarios: Acceleration.

Baseline: Chiara is a 20-year-old hairdresser. She still lives with her parents but is looking for a flat of her own. She loves to travel and works out regularly to stay in shape. Chiara is friends with Eleonore, who is studying at Harvard in the USA. Eleonore sometimes visits Chiara.

Acceleration scenario: Chiara works in a mobile and autonomous beauty salon. She spends four days a week in Paris and two days in Nice. On the last day of the week, she parties with her friends in a special bus. During her journeys between Paris and Nice, she likes to watch series on Netflix. Chiara is no longer looking for a flat. If she needs a place to stay overnight, she uses a self-contained mobile hotel. She doesn't need to go shopping anymore because she gets what she needs delivered in the autonomous vehicle. Eleonore takes a high-speed plane to visit Chiara in Nice. Chiara works out in a gym inside of an automated vehicle whilst commuting.

This section looked at people-centred design from different angles. After providing an overview of different related concepts, we presented a methodological framework to combine prospective work with persona-based approaches. The *Scenario Personanarrative* method was presented as well as examples of the workshop-based process and results. From this angle, we move on to the third category of design and decision-making methods.

3.3 Data-Driven Design and Decision-Making

Data on the mobility system is increasingly becoming available in cities across the globe. A great opportunity for a sustainable mobility transition is to make use of this data to take informed decisions. At the same time, there are challenges that need to be tackled in the scope of data-driven decision-making which will be discussed in the following sections. We focus on two ways of data-driven decision-making. First, mobility modelling and simulations are discussed. Afterwards, the focus shifts to the management of charging infrastructure, critical for the ongoing upscaling of EVs.

3.3.1 Modelling and Simulating Mobility Systems

Modelling mobility systems is strongly related to the term of *digital twins* that has been established over the past years. Whilst the understanding of such a system varies across scientific and application domains, Wildfire (2018) and Batty (2018) refer to 'reactive' and 'predictive' digital twins that allows to better define the decision-making aspect in the scope of this chapter. They see a digital twin as a representation of the real world with varying levels of interaction. In both cases, the technical system is able to observe the current system state (e.g., mobility patterns, mobility needs, infrastructure) and extrapolate to a future state, given specific inputs such as policies or technological and social changes. The idea is to obtain, through simulation and modelling, an idea of probable future states of the system under these conditions.

A 'reactive' digital twin operates on a short or operational timescale. In this category fall systems like intelligent traffic lights, lane management systems, or parking guidance systems. By fusing historical and current information, those systems try to estimate the future pressure on the system on a scale of minutes or hours and evaluate the mitigation effects of actions taken in the system. This way, intelligent short-term control of the system can permit to reduce congestion or emissions. The 'reactive' digital twin is inherently linked to the real system by providing inputs to it. It has, hence, been argued that a distinction between the 'real world' and the 'digital twin' can become increasingly difficult. Today, reactive digital twins are deployable solutions that can improve the system in the short term.

On the larger scale, one could view the process of identifying problems in the real world, performing scientific analysis and research, and passing on to political action and implementation as a reactive loop that is constantly faced with a changing environment. This process can also be supported by digital tools, which, however, are not autonomously acting agents in the system. Rather, the simulations and models *predict* system states given hypothetical inputs that then allow humans to act upon the results. A 'predictive' digital twin, hence, helps to make long-term decisions by evaluating numerically different potential futures. In the case of urban and mobility planning, many factors affect the evolution of the system including peoples' behaviour, infrastructure development, energy systems, or housing policies. The modelled systems are, hence, complex and cannot be understood using human back-of-the-envelope calculations or logical reasons. The great value of the respective modelling tools lies in their ability to capture the cumulative effects taking into account multiple complex interdependencies between system elements. They are able to capture non-linear outcomes and rebound effects.

The following sections will focus on making use of data for strategic decision-making in the sense of a predictive digital twin: First, data availability and provision is discussed, followed by an introduction to modern models of the transport system.

Data availability and provision

To date, no standardised system exists that is able to predict and model the mobility system of a city or region in a generic way. Whilst the available computational resources are a limiting factor, data availability may be the biggest challenge in setting up meaningful simulations of the mobility system. Common data sets that are required to understand human mobility are described in the following paragraphs.

Census data sets describe the population of a city or a territory. They may contain information on a zonal level indicating the frequency of certain household types (e.g., single, two persons with/without children) or provide individual samples of households or people that are found inside of these zones. The smaller the zones and the more information on the inhabitants (e.g., age, gender, educational status, income, car availability), the more detailed simulations can be set up.

Commuting data sets describe the major flows in an area which are the movements from and to work or the main educational activities of the population. They are often given in the form of origin–destination (OD) matrices indicating the number of people commuting from any zone A to any other zone A' in the study area, for instance, on the level of municipalities.

Household travel surveys (HTS) are data sets that are obtained by surveying the population on their daily mobility behaviour. Usually, these data sets cover the mobility of one reference person or, more rarely, all persons in a household. They describe in detail at which time the persons have performed activities like staying at home, being at work or school, going shopping during the day. Furthermore, there is information on where these activities have happened, and with which means of transport people have moved between these activities. These data can then be used to develop digital representations of the mobility of the population. Given a detailed HTS, other data sets, such as commuting matrices, can be obtained.

To further qualify why certain daily movements take place, often additional economic data is used. These may be enterprise census data indicating the employment distribution in the territory or income surveys that allow to model mobility in function of households' or municipalities' wealth.

Additionally, supply side data is needed for detailed transport simulations. These are mainly road network data describing, which today are ubiquitously available, for instance on OpenStreetMap, and public transport schedule data, often in the GTFS or the newer NeTEx format.[1] Using this data, the available modes of transport can be modelled in detail and their use by the population with their mobility needs can be evaluated.

To verify the quality of models, validation data is necessary. Traditionally, flow data has been the most important validation source indicating how many vehicles pass a specific point in a road network per day or even per hour. Similar data sets are available from transit operators per station or line in cities where validation of tickets or automatic counting of passengers is applied. Finally, new data sources such as zone-ton-zone travel times from services such as Uber[2] have become available as part of the endeavour of cities to encourage transport providers to share their data. Another active track of research focuses on the use of ubiquitous data such as GPS and mobile phone traces to increase the level of detail and robustness of transport simulations. A major question that arises is who the actors around the provision of mobility data are and what are their obligations. In general, the data provision process can be analysed in various steps:

1. *Data collection* describes the process of obtaining data. In some cases, like mobile phone data, the producer of the data sets is obvious as the mobile phone providers are the once responsible for collecting the data sets. In other cases, such as survey data, the situation is different. Whilst population data is often collected by national statistical offices, detailed mobility and HTS are performed on the local or regional level. Often, these surveys are performed by entities that have specialised on the recruitment of respondents, and the design and implementation of the surveys.

[1] https://netex-cen.eu [accessed August 2023].

[2] Uber Movements, https://movement.uber.com/ [accessed August 2023].

2. Next, *data processing* takes place. Often, raw data is not directly published to be used by the public or entities that are given access, but the information is modified to transform them into a shareable state. On the one hand, data needs to be *cleaned* because codes and identifiers are used that are hard to understand and work with for third parties. Data outliers may need to be detected and removed. On the other hand, *privacy issues* may need to be tackled because companies are obligated to not share personal information, especially in the light of recent legislation like the GDPR. These steps may pose a major challenge in the process of making the data useable, especially if the final goal is to publish them as open data to the public. Whilst some entities, dependent on legislation, may decide to not perform these activities and share raw data with trustable end users under specific data agreements, other may refrain from passing on any data. A major technical challenge remains to develop the protocols and tools to help entities transforming their data sets into a shareable format, legally and quality-wise.
3. Finally, *data publication and maintenance* remain an important challenge. Once the data has been acquired, do potential users need to contact the collecting entities, or are they generally available on a centralised platform? If so, a central entity needs to be responsible for maintaining access to the data sets. Such endeavours are fostered, for instance, in France by the establishment of a national data portal, supported by individual portals of the regions and metropolitan regions. On the European level, researchers and data providers are encouraged to follow the FAIR principles and open data repositories such as Zenodo.

The process of making data available such that it can be used either openly or by specific users, hence, faces multiple challenges. They are increasingly solved by public entities and companies following enabling legislation and encouragement by funding agencies, but the situation still varies greatly between different countries and regions. It is important to remember the challenges mentioned above when thinking of performing meaningful data-driven decision-making as the available data sets are the most important foundation of setting up the relevant tools. How the data can be used to set up the respective simulations and models is covered in the following section.

Mobility system modelling

Setting up a simulation of a city's transport system and drawing decisions from it is a complex task. It requires the availability of various types of data and usually applies a chain of interconnected models representing different aspects of the system. The standard reasoning behind transport models is that there is the *demand* that represents the mobility needs of the population and the *supply* which describes the infrastructure and the offered services to fulfil these mobility needs. Most strategic (predictive) simulations then apply a feedback loop in which the *supply* is evaluated against the *demand*, i.e., a decision-making process is simulated that resembles how people would likely make use of the available means of transport. Having established such a simulation loop, the task of the model developer is to *calibrate* the model so that relevant system indicators represent the reference situation adequately. This comparison provides evidence that the model works correctly and is able to generally reflect the functioning of the real system. Finally, the model user can define scenarios by introducing new technologies, policies, or behavioural changes into the system.

Demand modelling

A demand model presents the mobility needs of the population. The level of detail of the demand model strongly depends on the scope and goal of the decision-making process. Traditionally, transport authorities and planning agencies have divided cities into *Traffic Analysis Zones* (TAZ) that represent a nearly uniform distribution of daily trip origins. Hence, inside a TAZ, at any point, there is the same probability of observing somebody starting a trip using any means of transport, whilst it differs from another TAZ. Using a demand model, one can then establish a flow between two TAZ indicating, for instance, how many people commute every morning from one zone to another to go to work. The challenge in this process is to represent well the real flows between these zones in a model. In the framework of classic *four step models*, the first two modelling steps represent *Trip Generation* (deciding how many trips originate from which zone) and *Trip Distribution* (deciding how many of the originating trips of one zone go to any other zone). Trip generation is usually strongly linked to census data or information on the working population in a zone. Trip distribution, on the other hand, is often linked to economic activity and employment in the destination zones. The resulting data sets can be imagined as a quadratic table indicating the number of travellers from any origin to any destination zone. Using these data sets, modellers have traditionally been able to modify the flow relations to represent the development of new city districts or business parks. In the downstream models, it is then possible to, for instance, dimension the road and rail infrastructure to accommodate the new flows.

Modern transport planning has progressed to a more detailed representation of the mobility demand in a city. Individual-based models are used which represent in detail the households, persons, and their daily trips. The term *synthetic population* describes a data set that describes potentially millions of households in a region individually with specific attributes (such as the number of available cars and household income) and their individual members (defined by, e.g., age, gender, education). The aim of generating such a data set is to statistically represent the current population of a study area as closely as possible. Definitions of what is a 'good' synthetic population are still being developed, but commonly the observed frequency of age groups, gender, and other attributes or their combinations are used as quality criteria. Generating such data sets can be easy if sufficient information, for instance from detailed population census data sets, is available, or more challenging if only a small sample of information is available. Depending on the data availability, various approaches have been proposed from simple Statistical Weighting (Durán-Heras et al. 2018; Müller 2017; Yameogo et al. 2021) approaches using Machine Learning models like Bayesian networks (Sun and Erath 2015) and Hidden Markov Chains (Saadi et al. 2016), and, recently, Generative AI (Borysov et al. 2019).

Box 3.3 Synthetic demand for Île-de-France [method]

A recent example for a comprehensive process to generate a synthetic travel demand data set has been developed for the Île-de-France region around Paris (Hörl and Balac 2021a). It is a data set that can be generated by anybody as the process is replicable and entirely based on open data that is available in France. The process follows various steps to arrive at the final output.

First, the approach makes use of a detailed population census data set that is available as open data in France. Using this data set, it is possible to generate 12 million different persons within their households representing the whole population of the Île-de-France region. Whilst, at this point, their home location is known only on the level of statistical zones, a detailed building registry is used to define the specific home location of each household. The selection of buildings is based on the number of flats that is also indicated in the open data set. In the third step, a structural activity chain is attached to each person. For that, an open data national HTS is used. The survey contains the daily activity schedules and socio-demographic attributes of more than 14,000 persons for the study area. Using the sociographic attributes, one activity chain including the start and end times of activities from the HTS is selected for each synthetic person. Finally, the locations of all assigned activities are chosen based on a novel assignment process (Hörl and Axhausen 2021).

The result of the process is a *synthetic travel demand* data set that describes in detail (by second, by coordinate) the mobility demand of the population Paris. Since all used data sets and the code are open and publicly available (Hörl and Balac 2021b), the process is fully replicable. Furthermore, the used data sets are available for all of France, which has led to various reuses and adaptations, for instance, for Nantes (Le Bescond et al. 2021), Rennes (Leblond et al. 2020), Lyon (Hörl and Puchinger 2022), and Lille (Diallo et al. 2021). Beyond Europe, the process has also been adapted to California (Balac and Hörl 2021) and Sao Paulo (Sallard et al. 2021), where similar open data sets as in France are available.

After establishing a *synthetic population*, each person receives a daily activity schedule, indicating when and where a person wants to perform an activity (e.g., staying at home, going to work, shopping, going to school). Adding this information leads to a *synthetic travel demand* data set which is able to describe the mobility of the population by examining the trips that are necessary to move from one activity to the next

(Hörl and Balac 2021a). The most common approach of attaching activity chains to individual persons is to make use of a HTS and a Statistical Matching method (Namazi-Rad et al. 2017). In this approach, a data set of daily mobility chains, linked to socio-demographic information is already available. The Statistical Matching procedure then searches for every person in the *synthetic population,* a chain in the HTS data set that resembles the person's socio-demographic attributes as closely as possible. Other approaches are to generate activity chains from scratch, for instance, using Bayesian networks (Joubert and de Waal 2020). The *synthetic demand* data set is then finalised by assigning distinct locations in the study area to each activity in the person schedules (Hörl and Axhausen 2021). An interesting current thread of research is to make use of (anonymised) phone trajectory data to generate daily activity schedules (Anda et al. 2021; Ziemke et al. 2021).

In summary, detailed *synthetic demand* data sets allow for a detailed analysis and simulation of the daily mobility of the population. They follow the paradigm of activity-based modelling in which the activities that people intend to perform, during the day, are seen as the source of mobility, rather than performing a bare phenomenological analysis such as in the classic flow-based models described above. Furthermore, they allow analysing mobility patterns by time of day rather than looking at specific time slices (morning peak, evening peak, off-peak). Finally, because they do not represent trip origin and destinations on the level of aggregated zones, but individual locations, they also allow for a detailed analysis of spatially detailed and dynamic mobility services.

Behavioural modelling

Once the daily mobility demand in a region has been established, the short-term behaviour of the population needs to be modelled. The most common assumption when modelling (average) mobility behaviour is to assume that travellers are rational decision-makers—*homo economics*. Faced with a range of alternatives (for instance, taking the bus or the car), they based their decision on various choice dimensions such as the expected cost, travel time, and number of transfers. Such decisions are most often modelled by models from discrete choice theory such as the Multinomial Logit Model (Train 2009). Given the respective choice properties of all alternatives (e.g., travel times, costs, comfort), these models provide a probability of a decision-makers preferring one or the other

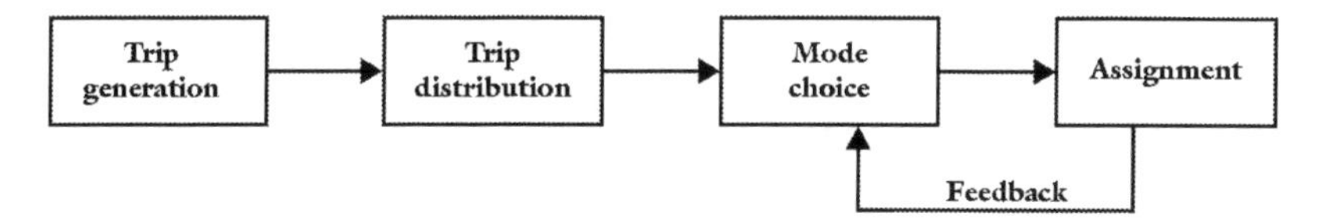

Fig. 3.3 Phases of the four-step model

option. Hence, the models can be used to simulate the travel decisions of the population.

In the classic *four-step model* approach, trip generation and trip distribution is followed by the *mode choice* step (Fig. 3.3). At this point, the models evaluate, for each flow (from zone A to zone B), the travel characteristics and obtain which mode of transport (car vs. transit) would be used by how many of the travellers. The models are, hence, sensitive to the observed travel times. Implementing, for instance, a new subway line into the area will reduce travel times for certain travel relations and, hence, increase the attraction of public transport. This way, modellers may test infrastructure investments and their effect on mobility patterns.

In more disaggregated models that are based on activities and synthetic travel demand data sets, more decisions may be simulated. First, departure times may be modelled. Since not only commuting flows at specific times of the day are simulated, agents may decide to depart earlier or later for their individual trips during the day. This way, they may adapt to opening times of shops or leisure facilities or adapt better to their work schedule. Likewise, agents may make decisions for the places at which activities are performed, for instance, if a new gym is opened that is more convenient to reach from a person's home or work location. In general, as described above, dynamic agent-based and activity-based simulations allow for a much more detailed representation, not only of the mobility demand, but also how mobility decisions are made.

Feedback and system equilibrium

Complexity is introduced into mobility simulations because usually the factors on which decisions are made by the traveling population are, in turn, impacted by their decisions. If the whole population suddenly decides to switch to individual traffic using their own car, congestion may increase strongly, rendering the existing public transport connections more favourable in terms of travel time. Hence, some persons may switch back to public transport because they see it as the better alternative for their trips. As persons switch to the public transport system, congestion and travel times decrease, but at the same time, crowding in the trains and buses may rise. Eventually, a sufficient number of persons may use the public transport system and the road infrastructure that the trade-off between travel times, costs, comfort, and other dimensions between the two modes of transport reach an equilibrium. In other words, the system has stabilised into a situation in which all decision-makers are satisfied with their decision. Unilaterally, they cannot perform any decision that will improve their individual situation, a state which is also referred to as the user equilibrium (UE). In most mobility modellers, the real-world state of the system at a given point in time is hypothesised to be in such a UE state that should be recovered by a simulation to make sure that the reference situation is modelled properly.

The last step of the four-step model is called the *assignment step*. At this point, all agent decisions are evaluated and loaded onto the road and public transport networks. This means that if one zone can be reached by another one through two different routes, but with the same travel time, half of the travellers should use route A, and half of them route B. Since the travel time on any route is affected by the number of travellers, an equilibrium also needs to be found on the route or route assignment level. Once this assignment is found, the four-step model has obtained the congestion levels and travel times throughout the study area. The obtained travel times can then be fed back to the *Mode choice* step to make more informed decisions.

Likewise, agent-based simulations are often performed in an iterative way. A share of the individual persons makes decisions (e.g., on departure times, transport modes, activity locations) and then the system is simulated, taking into account all the interactions and capacity restrictions in the system. Hence, new travel times and trip characteristics emerge, which, in turn, affect other agents in the next iteration of the simulation. These steps are repeated until an equilibrium is reached. This equilibrium may be defined in various ways, for instance, in terms of the overall percentage of trips that are performed with one or another mode of transport (Hörl et al. 2021).

The process described above explains the complexity of transport and mobility simulations, both in classic four-step models and more recent agent-based models. Usually, to reach the desired equilibrium state, hundreds or thousands of iterations are required in which thousands or millions of decisions need to be simulated. The computational effort to perform transport simulations for a whole city or region, therefore, remains substantial.

Calibration and validation

The high-computational effort of transport models especially poses a challenge for their calibration. It describes the process of finding the adequate model parameters to correctly represent the mobility system like one can observe in reality. This reference state must be qualified by reference data. These may be the observed mode shares (percentage of mode use) on a daily basis that has been obtained through surveys, it may be the number of vehicles counted at multiple intersections throughout the city, or the number of ticket tap-ins in a subway station. The modeller needs to define the calibration metrics and the level of allowed difference that is deemed as 'correct'. Parameters that may be adapted in the simulation include the size and allowed speeds in the road network, the weighting of different choice dimensions (travel time vs. cost) in the behavioural model of the decision-makers, along many others.

For four-step models, detailed official guidelines on the calibration quality are available, for instance for London. To date, no commonly applied guidelines exist for more detailed agent-based simulations.

Variant development

Once a reliable baseline simulation has been set up, changes can be introduced to the system. For instance, a modeller might add a new lane to an existing road or introduce a new road to the system. Likewise, a new subway line may be added. Using the simulation and the well-calibrated baseline situation, the modeller can then obtain indicators such as the travel time between and origin and a destination and verify if it would likely improve and by which magnitude. Such analyses would be possible using classic four-step models.

On the other hand, the trend in sustainable urban mobility is to limit the addition of new infrastructure but make use of the existing built environment. This is reflected by the increasing use of digital technology and ubiquitous availability of mobile devices that enable the development of ride-sharing and on-demand services. In the ideal case, these systems improve the shared use of the infrastructure by matching in detail supply and demand with a high dynamism and spatial granularity. To understand these new services, however, it is necessary to also be able to represent them adequately in the simulation. This is only possible using the detailed representation of mobility demand and decision-making that agent-based simulations offer.

Box 3.4 Modelling the local impact of the Grand Paris Express [method]

The synthetic population generated by the process presented in the previous box describes the travel demand for the area of Île-de-France. It can be combined with supply related data (road network provided by OpenStreetMap and public transit schedules provided by Île-de-France Mobilités in GTFS format) to build large-scale simulations of the mobility of the area. The calibrated choice model enables to reproduce how travellers chose between mode alternatives according to, e.g., the cost, travel time, and wait time implied by each mode. The calibration of the choice models ensures that when the simulation is performed with the current offer, global behaviours similar to the real world ones are observed.

Building upon a reference simulation that is able to reproduce the current mobility phenomena, prospective simulations can be performed in order to assess the impact that changes on the demand or on the offer side would have on the overall mobility. In most studies, the effect of evolutions of the offer is investigated. A wide range of mobility solutions are assessed, designed, and dimensioned using agent-based simulations. We give here as an example the study first presented by Chouaki (2023).

In the area of Île-de-France, the public transport network is currently witnessing major developments. The best-known project is called the Grand Paris Express, which aims to create a new rapid transit system connecting the suburbs of Paris and reducing travel time and congestion. The project includes the construction of four new metro lines (15 to 18) and the extension of two existing ones (11 and 14), covering a distance of 200 kilometres and serving 68 stations, thus, doubling the current already very extensive metro network of Paris. Other tramway projects are also planned throughout the area and intend to further reinforce the public transport offer. The availability of data regarding the future lines and their operation (expected frequencies and travel times) allows to add them into the simulated offer relatively easily.

A prospective simulation with the future planned mobility supply is then constructed and performed. The use of a behavioural model allows to assess the potential of new lines to encourage modal shift from cars to public transport. Moreover, the redistribution of the traveller's demand in public transport is estimated in detail. The results presented suggest that new lines will mainly attract passengers from existing bus lines that are rendered less attractive (Chouaki 2023).

The study also sheds light on the lack of data regarding the evolution of bus lines around new stations of the high-capacity ones. Thus, resulting in an underestimation of the ability of travellers to access them. Future research ought to address this issue.

Accordingly, there is a large range of technological and policy interventions that have been analysed using agent-based transport simulation, ranging from car sharing services (Balac et al. 2017), Mobility-as-a-Service (Becker et al. 2020), impacts of automated vehicles (Hörl et al. 2021; Kaddoura et al. 2020), mitigation of rail interruptions (Leng and Corman 2020), assessment of dispatching algorithms for on-demand mobility (Hörl et al. 2019; Maciejewski et al. 2017), or the optimal design of transit lines (Manser et al. 2020). Whilst the previous examples are based on the Multi-Agent Transport Simulation (MATSim) framework, there are similar and alternative open-source simulation platforms such as SUMO (Lopez et al. 2018), SimMobility (Azevedo et al. 2017), POLARIS (Auld et al. 2016), and commercial options with different levels of accessibility and documentation.

3.3.2 Charging Management of EVs[3]

Data-driven approaches can help the design and planning of new solutions or infrastructure for urban mobility systems. They are also critical to manage and optimise sub-systems, such as the energy system responsible for the charging vehicles. As introduced in the sub-chapter on technological trends, a range of context-dependent challenges comes with the increasing market share of EVs. Therefore, this section aims to contribute to the topic of coordinated charging management of EVs by providing a brief analysis of the challenges related to EV-power system coordinated charging problem. Through our critical analysis, we provide a categorisation to understand this problem within four major axes, as depicted in Fig. 3.4. The division serves as a framework to examine all aspects of the EV charging coordination problem. The categories include the *Objective* category which identifies the possible main goals pursued to study the coordinated EV charging problem both from the 'user' perspective and the 'operators' perspective. The *Pricing Strategy* and *Charging Control Architecture* categories offer contextual information about the EV charging problem and their links to electric power systems management. Finally, the *Optimisation technique* category briefly outlines the most prominent mathematical modelling and numerical optimisation approaches employed to model and optimise optimal coordinated charging policies. This approach allows for an evaluation of the various strategies and techniques employed in tackling the coordinated charging management of EVs and is aimed to be useful for researchers and practitioners seeking to further investigate this problem.

The classification considers the different usage perspective, notably, user-based (EV owner) and operator-based (system operator) objectives. User-based objectives seek to improve EV drivers' welfare by minimising energy costs and reducing charging waiting times. On the other hand, operator-based objectives focus on optimising grid and power distribution systems, addressing challenges caused by uncoordinated EV charging, such as peak load reduction, grid overload minimisation, and voltage stability. Moreover, the section covers the pricing strategies for EV charging, emphasising dynamic pricing policies like Real-Time Pricing,

[3] The authors would like to acknowledge the contribution of Yassine Benider to this section as part of his master thesis work at CentraleSupélec under the supervision of Adam Abdin.

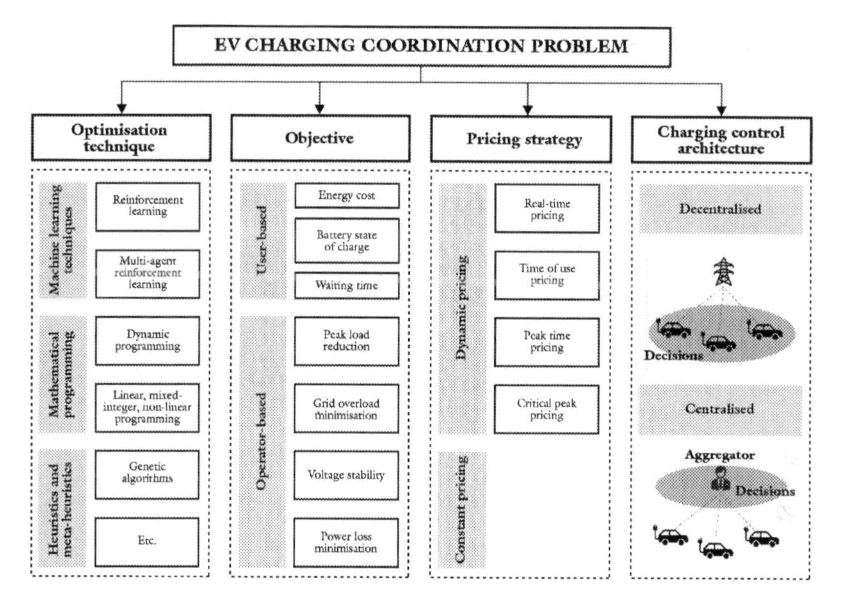

Fig. 3.4 Classification of coordinated EV charging management problems

Time of Use, and Critical Peak Pricing, which significantly influence customer electricity consumption and charging behaviour, playing a key role in the overall coordination of EV charging. The focus of the surveyed research work is to investigate the different modelling and solution methods proposed to treat this problem, such as multi-agent reinforcement learning, mathematical programming, game-theoretic approaches, and hybrid algorithms.

Objective of EV charging coordination

User-based objectives in EV charging coordination aim to prioritise the welfare of EV drivers. These objectives include minimising energy costs and reducing waiting times for charging. One of the most significant concerns addressed in the literature is the charging energy cost, which is a critical factor influencing EV adoption and penetration. Various studies propose solutions to tackle this challenge, such as single agent (Arif et al. 2016; Chiş et al. 2016) and multi-agent (Cao et al. 2012; Qian et al. 2022) Reinforcement Learning approaches aiming to reduce EV charging costs. Other research work proposes mathematical programming-based approaches, like linear programming and mixed integer nonlinear programming, with the objective of charging cost reduction (Bitencourt et al. 2017; Zhang et al. 2017). Additionally, heuristic and metaheuristics methods, such as the placement algorithm and particle swarm, are employed with the same objective (Usman et al. 2021; Suyono et al. 2019). Other research works incorporate renewable energy generator sources into their models to minimise energy costs (Yang et al. 2019). Besides energy costs and EV battery state of charge, other important considerations are the maximisation of photovoltaic self-consumption to treat photovoltaic generation as a source of energy (Dorokhova et al. 2021), and efforts to minimise greenhouse gas emissions from power generation (Zhang et al. 2017). Moreover, authors acknowledge the importance of addressing the delay in customer demand satisfaction, specifically the waiting time for desired charging (Qian et al. 2019; Ferro et al. 2018).

Operator-based objectives in EV charging coordination are centred around ensuring the efficient functioning of the grid and power distribution systems. These objectives address various critical issues resulting from uncoordinated EV charging, including peak load reduction, grid overload minimisation, voltage stability, and power loss reduction. Several papers propose solutions to achieve grid overload minimisation. Some of these solutions are based on multi-agent reinforcement learning, with the objective of reducing the stress on the grid (Da Silva 2020; Tuchnitz 2021). Mathematical programming-based approaches are also employed by other authors to achieve the same objective (Korolko and Sahinoglu, 2015; Dubey and Santoso 2015). Additionally, a game-theoretic approach is utilised in some studies to manage charging and address grid overload (Karfaopoulos and Hatziargyriou 2012; Hu et al. 2015; Cao and Chen 2018). Several methods are presented in the literature to control EV charging loads and lower demand during peak hours (Cao et al. 2012; Bitencourd et al. 2017; Chen et al. 2017). Other objectives pursued by research works include reducing voltage irregularities and enhancing voltage stability within the system (Korolko and Sahinoglu 2015; Soares et al. 2017) as well as reducing distribution network power losses (Suyono et al. 2019; Esmaili and Goldoust 2015). Moreover, certain studies focus on leveraging Photovoltaic generation as a source of energy to maximise

Photovoltaic self-consumption and alleviate stress on the grid (Dorokhova et al. 2021).

Charging control architectures

Another classification aspect identified in this analysis is based on the type of charging control architecture. Charging and discharging operations can be controlled either in a distributed manner (each vehicle independently) or centrally. We, thus, classify this problem into three primary structures: centralised, decentralised, and hybrid charging control architecture framework, combining elements from both centralised and decentralised approaches.

The *centralised charging control* architecture involves a local controller or aggregator responsible for coordinating the charging process of a group of EVs, ensuring benefits for both the network operator and EV users (Wang et al. 2014). However, this centralised architecture can be complex, limiting the number of EVs a single agent can handle effectively. Several papers present solutions within a centralised EV charging architecture with an EV fleet aggregator (Maigha and Crow 2014; Dimitrov and Lguensat 2014; Ruelens et al. 2012; Vandael et al. 2015). For example, one study by Maigha and Crow (2014) proposes a centralised scheduling strategy for PHEV charging using a genetic algorithm, showcasing how it enhances the load curve's shape. Another paper by Ruelens et al. (2012) introduces a stochastic method for the aggregator of a hybrid EV fleet to schedule their charging. Additionally, Dimitrov and Lguensat (2014) apply Q-learning to analyse the impact of EV scheduling on charging stations' revenues and maximise profitability. Notably, only a few papers (Arif et al. 2016; Tuchnitz et al. 2021; Vandael et al. 2015) within this review adopt single Reinforcement Learning approaches with a centralised architecture, as the majority of adopters prefer mathematical programming and heuristic solutions.

In the *decentralised charging control architecture*, individual EV users have autonomy in making decisions about their charging behaviour. However, without regulations from network aggregators, the best outcomes for the grid and distribution network may not always be guaranteed. Charging behaviours in this framework can be influenced through appealing incentives under dynamic electricity pricing. Unlike the centralised architecture, the decentralised approach offers greater flexibility, scalability, and ease of adoption for EV users, who can control the benefits they derive from their charging actions, including energy cost reduction, battery state of charge, and waiting time (Chiş et al. 2016; Wan et al. 2018; Li et al. 2019).

Hybrid charging control architectures are also proposed in the literature to address the challenges of coordinated EV charging (Wang et al. 2017). In this hybrid architecture, three distinct parts work in tandem. Firstly, the centralised component adopts an offline optimal scheduling approach with the primary objective of minimising energy costs. Secondly, the decentralised part models the interactions between EVs and the aggregator. Finally, the third part focuses on maximising the overall system goals by analysing how the two previous components impact the utility and efficiency of the charging process.

Pricing strategies for EV charging

Electricity pricing covers electricity generation, transmission, and distribution expenses (Soares et al. 2017). These costs are charged to EV consumers within a charging tariff. Pricing policies are set by electricity suppliers, typically charged per kilowatt-hour (kWh), and can vary widely between suppliers and countries. Traditional flat tariffs, block rate tariffs, and other static pricing schemes are deemed insufficient for modern smart grids and intelligent charging devices such as EVs (Yang et al. 2019).

The electricity consumption behaviour of end users can be influenced by demand response programs introduced by electricity providers. These programs prompt consumers to alter their electricity usage patterns in response to changing electricity prices or incentives aimed at reducing consumption during peak periods or ensuring network reliability (Soares et al. 2017).

Regarding EV charging coordination, the focus is on dynamic pricing strategies, as constant pricing policies do not incentivise users to minimise charging costs. Dynamic Pricing Strategies can significantly impact customer electricity consumption and charging behaviour. Several dynamic pricing policies are defined, including Real-Time Pricing, Time of Use, Peak Time Rebate, and Critical Peak Pricing. These policies play a crucial role in managing EV charging behaviour efficiently and effectively.

Real-Time Pricing is a charging policy characterised by variable rates that are adjusted regularly, typically every hour or a few minutes, to reflect real-time changes in electricity demand (Korolko and Sahinoglu 2015). Under this policy, the cost of electricity follows the actual demand, making it advantageous for electricity suppliers but challenging for domestic users to follow due to the constantly changing prices.

Several studies (Hu et al. 2015; Cao and Chen 2018; Wang et al. 2020) have utilised the Real-Time Pricing policy to simulate their solutions for EV charging coordination. For instance, Hu et al. (2015) and Cao and Chen (2018) employ the Real-Time Pricing scheme in their proposed models, considering EV charging coordination as a non-cooperative game governed by a market-based multi-agent system.

The *Time of Use* policy involves dividing the day into a number of periods based on electricity demand and time, and to price each period according to predicted information of the usage patterns. A typical Time of Use pricing divides the day into peak hours, off-peak hours, and mid-peak hours. Peak hours typically occur in the late afternoon and evening, when energy demand is highest due to people returning home from work. During these peak hours, local energy suppliers charge the highest electricity rates. The Time of Use prices are announced in advance and are determined based on historical data, unlike Real-Time Pricing, which reflects real-time changes in demand.

Some studies have utilised the Time of Use pricing scheme to validate their models in real-world scenarios (Usman et al. 2021; Maigha and Crow 2014). For example, Usman et al. (2021) propose an optimal charging starting time-based scheduling method that considers the departure time of electric vehicles. They organise the night-time charging event in a way that minimises network losses and allows EV users to benefit from the cost-effective rate zones offered by the Time of Use scheme.

Critical Peak Pricing is a variant of Time of Use pricing, specifically designed to address high-peak demand periods by implementing electricity pricing adjustments during these critical times (Wang and Li 2011). Whilst Time of Use pricing applies different rates for electricity consumption based on historical data, Critical peak pricing takes into account real-time information and forecasts of high-demand periods to set pricing (Amin et al. 2020). Unlike Time of Use, Critical peak pricing reacts swiftly and applies higher prices during peak hours to effectively reduce peak loads and alleviate stress on the grid (Newsham and Bowker 2010).

Although Critical Peak Pricing schemes were less prevalent in the literature review, some authors have explored their potential benefits. For instance, Yin et al. (2015) proposed a Critical peak pricing optimisation model to address peak load issues. Similarly, IDusparic et al. (2013) introduced a distributed Multi-agent Reinforcement Learning approach for EV charging management, incorporating load forecasting for residential demand within a Critical peak pricing context.

Overall, Critical peak pricing offers an alternative approach to dynamic pricing, demonstrating its potential effectiveness in managing peak loads during critical periods of high-electricity demand. Additionally, various other pricing strategies have been explored in the context of EV charging coordination, each with its unique advantages and challenges. The research in this area remains ongoing, with an increasing focus on finding optimal pricing policies to strike a balance between user-based objectives and grid optimisation, fostering the widespread adoption of electric vehicles and promoting a sustainable future for energy systems.

In this section, we presented an overview and a brief analysis of various methods employed to address the challenges arising from uncoordinated EV charging as it relates to power systems operations. Our analysis explored different real-world contexts, including energy tariffs and charging control architectures, and identified the importance of compatibility with existing utility tariffs for successful implementation. Future research areas for dealing with real-world challenges in EV charging management include optimising communication between EVs and the grid, assessing the impact of charging management schemes on EV batteries and energy needs, and addressing the challenges of integrating EV charging into smart grids whilst considering cost breakdowns for software and hardware implementation.

References

European Commission (EC). (2011) Roadmap to a single European transport area—Towards a competitive and resource efficient transport system. EU white paper COM(2011) 144. Brussels: European Commission.

Al Maghraoui, O., Vallet, F., Puchinger, J., and Yannou, B. (2019) Modeling traveler experience for designing urban mobility systems. *Design Science*, Vol. 5/E7. https://doi.org/10.1017/dsj.2019.6

Amin, A., Tareen, W.U.K., Usman, M., Ali, H., Bari, I., Horan, B., Mekhilef, S., Asif, M., Ahmed, S., and Mahmood, A. (2020) A review of optimal charging strategy for electric vehicles under dynamic pricing schemes in the distribution charging network. *Sustainability*, Vol. 12, p. 10160. https://doi.org/10.3390/su122310160

Anda, C., Ordonez Medina, S.A., and Axhausen, K.W. (2021) Synthesising digital twin travellers: Individual travel demand from aggregated mobile phone data. *Transportation Research Part C: Emerging Technologies*, Vol. 128, p. 103118. https://doi.org/10.1016/j.trc.2021.103118

Arif, A.I., Babar, M., Imthias Ahamed, T.P., Al-Ammar, E.A., Nguyen, P.H., René Kamphuis, I.G., and Malik, N.H. (2016) Online scheduling of plug-in vehicles in dynamic pricing schemes. *Sustainable Energy, Grids and Networks*, Vol. 7, pp. 25–36, https://doi.org/10.1016/j.segan.2016.05.001.

Auld, J., Hope, M., Ley, H., Sokolov, V., Xu, B., and Zhang, K. (2016) POLARIS: Agent-based modeling framework development and implementation for integrated travel demand and network and operations simulations. *Transportation Research Part C: Emerging Technologies*, Vol. 64, pp. 101–116. https://doi.org/10.1016/j.trc.2015.07.017

Azevedo, C.L., Deshmukh, N.M., Marimuthu, B., Oh, S., Marczuk, K., Soh, H., Basak, K., Toledo, T., Peh, L.-S., and Ben-Akiva, M.E. (2017) Simmobility short-term: An integrated microscopic mobility simulator. *Transportation Research Record Journal of the Transportation Research Board*, Vol. 2622, pp. 13–23. https://doi.org/10.3141/2622-02

Balac, M., and Hörl, S. (2021) Synthetic population for the state of California based on open-data: examples of San Francisco Bay area and San Diego County. In 100th Annual Meeting of the Transportation Research Board. Washington, D.C., January 2021.

Balac, M., Ciari, F., and Axhausen, K.W. (2017) Modeling the impact of parking price policy on free-floating carsharing: Case study for Zurich, Switzerland. *Transportation Research Part C: Emerging Technologies*, Vol 77, pp. 207–225. https://doi.org/10.1016/j.trc.2017.01.022

Batty, M. (2018) Digital twins. *Environment and Planning B: Urban Analytics and City Science*, Vol. 45, pp. 817–820. https://doi.org/10.1177/2399808318796416

Becker, H., Balac, M., Ciari, F., and Axhausen, K.W. (2020) Assessing the welfare impacts of Shared Mobility and Mobility as a Service (MaaS). *Transportation Research, Part A: Policy and Practice*, Vol. 131, pp. 228–243. https://doi.org/10.1016/j.tra.2019.09.027

Bertolini, L. (2020) From "streets for traffic" to "streets for people": Can street experiments transform urban mobility? *Transport Reviews*, pp. 1–20. https://doi.org/10.1080/01441647.2020.1761907.

Bertram, B., and Berthold, B.F. (2012) Was Sind Sinus-Milieus®? In: Thomas, P.M. and Calmbach, M. (eds.), *Jugendliche Lebenswelten: Perspektiven für Politik, Pädagogik und Gesellschaft*. Berlin, Heidelberg: Springer Berlin Heidelberg, pp. 11–35.

Cao, T. et al. (2012) An Optimized EV Charging Model Considering TOU Price and SOC Curve. IEEE Transactions on Smart Grid, Vol. 3/1, pp. 388–393. https://doi.org/10.1109/TSG.2011.2159630

De Bitencourt, L.A., Borba, B.S.M.C., Maciel, R.S., Fortes, M.Z., and Ferreira, V.H. (2017) Optimal EV charging and discharging control considering dynamic pricing. 2017 IEEE Manchester PowerTech, Manchester, UK, 2017, pp. 1–6. https://doi.org/10.1109/PTC.2017.7981231

Börjeson, L., Höjer, M., Dreborg, K.-H., Ekvall, T., and Finnveden, G. (2006) Scenario types and techniques: Towards a user's guide. *Futures*, Vol. 38/7, pp. 723–739. https://doi.org/10.1016/j.futures.2005.12.002

Bornet, C., and Brangier, É. (2013) La méthode des personas: principes, intérêts et limites, *Bulletin de psychologie*, Vol. 524/2. https://doi.org/10.3917/bupsy.524.0115/

Borysov, S.S., Rich, J., and Pereira, F.C. (2019) How to generate micro-agents? A deep generative modeling approach to population synthesis. *Transportation Research Part C: Emerging Technologies*, Vol. 106, pp. 73–97. https://doi.org/10.1016/j.trc.2019.07.006

Buur, J. and Matthews, B. (2008) Participatory innovation. *International Journal of Innovation Management*, Vol. 12/3, pp. 255–273. https://doi.org/10.1142/S1363919608001996

Cao, C., and Chen, B. (2018) Generalized Nash equilibrium problem based electric vehicle charging management in distribution networks. *International Journal of Energy Research*, Vol. 42, pp. 4584–4596. https://doi.org/10.1002/er.4194

Charreaux, G. (2004) Les théories de la gouvernance: de la gouvernance des entreprises à la gouvernance des systèmes nationaux (No. 1040101). Université de Bourgogne-CREGO EA7317 Centre de recherches en gestion des organisations.

Chen, Q. et al. (2017) Dynamic price vector formation model-based automatic demand response strategy for PV-assisted EV charging stations. *IEEE Transactions on Smart Grid*, Vol. 8/6, pp. 2903–2915. https://doi.org/10.1109/TSG.2017.2693121

Chiş, A., Lundén, J., and Koivunen, V. (2016) Reinforcement learning-based plug-in electric vehicle charging with forecasted price. *IEEE Transactions on Vehicular Technology*, Vol. 66/5, pp. 3674–3684.

Chouaki, T. (2023). Agent-based simulations of intermodal mobility-on-demand systems operated by reinforcement learning. Doctoral dissertation, University Paris-Saclay.

Cooper, A. (1999) *The inmates are running the asylum*. New York: Macmillan.

Courmont, A., and Le Galès, P. (2019) Gouverner la ville numérique. Ville et numérique, Presses Universitaires de France. 9782130815259.

Da Silva, F.L., Nishida, C.E.H., Roijers, D.M. and Costa, A.H.R. (2020) Coordination of Electric Vehicle Charging Through Multiagent Reinforcement Learning, *IEEE Transactions on Smart Grid*, Vol. 11/3, pp. 2347–2356. https://doi.org/10.1109/TSG.2019.2952331/

Diallo, A.O., Doniec, A., Lozenguez, G., and Mandiau, R. (2021) Agent-based simulation from anonymized data: An application to Lille metropolis. *Procedia Computer Science*, Vol. 184, pp. 164–171. https://doi.org/10.1016/j.procs.2021.03.027

Dimitrov, S., and Lguensat, R. (2014) Reinforcement learning based algorithm for the maximization of EV charging station revenue. 2014 International Conference on Mathematics and Computers in Sciences and in Industry. IEEE.

Docherty, I., Marsden, G., and Anable, J. (2018) The governance of smart mobility. *Transportation Research Part A: Policy and Practice*, Vol. 115, pp. 114–125. https://doi.org/10.1016/j.tra.2017.09.012

Dorokhova, M., Martinson, Y., Ballif, C., and Wyrsch, N. (2021) Deep reinforcement learning control of electric vehicle charging in the presence of photovoltaic generation. *Applied Energy*, Elsevier, Vol. 301(C). https://doi.org/10.1016/j.apenergy.2021.117504.

Dubey, A., and Santoso, S. (2015) Electric vehicle charging on residential distribution systems: Impacts and mitigations. *IEEE Access*, Vol. 3, pp. 1871–1893. https://doi.org/10.1109/ACCESS.2015.2476996

Durán-Heras, A., García-Gutiérrez, I., and Castilla-Alcalá, G. (2018) Comparison of iterative proportional fitting and simulated annealing as synthetic population generation techniques: Importance of the rounding method. *Computers Environment and Urban System*, Vol. 68, pp. 78–88. https://doi.org/10.1016/j.compenvurbsys.2017.11.001

IDusparic, I., Harris, C., Marinescu, A., Cahill, V., and Clarke, S. (2013) Multi-agent residential demand response based on load forecasting. 2013 1st IEEE Conference on Technologies for Sustainability (SusTech), Portland, OR, USA, pp. 90–96. https://doi.org/10.1109/SusTech.2013.6617303.

Dyson, P., and Sutherland, R. (2021) *Transport for humans. Are we nearly there yet?* London Publishing Partnership.

European Commission (EC). (2021) Delivering the European Green Deal. https://commission.europa.eu/strategy-and-policy/priorities-2019-2024/european-green-deal/delivering-european-green-deal [accessed August 2023].

Elioth, Egis Group. (2017) Paris, an air of change. Towards carbon neutrality in 2050. https://paris2050.elioth.com/en/ [accessed August 2023].

Esmaili, M., and Goldoust, A. (2015) Multi-objective optimal charging of plug-in electric vehicles in unbalanced distribution networks. International *Journal of Electrical Power & Energy Systems*, Vol. 73, pp. 644–652. https://doi.org/10.1016/j.ijepes.2015.06.001

Fergnani, A., and Jackson, M. (2019) Extracting scenario archetypes: A quantitative text analysis of documents about the future, *Future & Foresight Science*, Vol. 1/2, pp. 1–14. https://doi.org/10.1002/ffo2.17

Fergnani, A. (2019) The future persona: A futures method to let your scenarios come to life. *Foresight*. https://doi.org/10.1108/FS-10-2018-0086.

Ferro, G., Laureri, F., Minciardi, R., and Robba, M. (2018) An optimization model for electrical vehicles scheduling in a smart grid. *Sustainable Energy, Grids and Networks*, Vol. 14, pp. 62–70. https://doi.org/10.1016/j.segan.2018.04.002

Fogg, B.J. (2009) A behavior model for persuasive design. In Proceedings of the 4th International Conference on Persuasive Technology (Persuasive'09). Association for Computing Machinery, New York, NY, USA, Article 40, pp. 1–7. https://doi.org/10.1145/1541948.1541999.

Fuglerud, K.S., Schulz, T., Janson, A.L., and Moen, A. (2020) Co-creating persona scenarios with diverse users enriching inclusive design. In: Antona M., Stephanidis C. (eds.), *Universal access in human-computer interaction*. Cham: Springer. https://doi.org/10.1007/978-3-030-49282-3_4.

Gall, T., and Haxhija, S. (2020) Storytelling of and for planning: Urban planning through participatory narrative-building, 56th ISOCARP World Planning Congress, Doha/online, 8 November 2020–4 February 2021.

Gall, T., Vallet, F., Douzou, S., & Yannou, B. (2021) Re-defining the system boundaries of human-centred design. *Proceedings of the Design Society*, Vol. 1, pp. 2521–2530. https://doi.org/10.1017/pds.2021.513

Gehl, J. (2011) *Life between buildings: Using public space*, 6th ed. London: Island Press.

Gregory, J. (2003) Scandinavian approaches to participatory design. *International Journal on Engineering Education*, Vol. 19/1, pp. 62–74.

Hörl, S. and Axhausen, K. W. (2021) Relaxation–discretization algorithm for spatially constrained secondary location assignment. *Transportmetrica A: Transport Science*, Vol. 19/2, pp. 1–20. https://doi.org/10.1080/23249935.2021.1982068

Hörl, S., and Balac, M. (2021a) Synthetic population and travel demand for Paris and Île-de-France based on open and publicly available data. *Transportation Research Part C: Emerging Technologies*, Vol. 130, p. 103291. https://doi.org/10.1016/j.trc.2021.103291

Hörl, S., and Balac, M. (2021b) Open synthetic travel demand for Paris and Île-de-France: Inputs and output data. *Data Brief*, Vol. 39, p. 107622. https://doi.org/10.1016/j.dib.2021.107622

Hörl, S., and Puchinger, J. (2022) From synthetic population to parcel demand: Modeling pipeline and case study for last-mile deliveries in Lyon. Presented at the Transport Research Arena (TRA) 2022, Lisbon.

Hörl, S., Ruch, C., Becker, F., Frazzoli, E., and Axhausen, K.W. (2019) Fleet operational policies for automated mobility: A simulation assessment for Zurich. *Transportation Research Part C: Emerging Technologies*, Vol. 102, pp. 20–31. https://doi.org/10.1016/j.trc.2019.02.020

Hörl, S., Becker, F., and Axhausen, K.W. (2021) Simulation of price, customer behaviour and system impact for a cost-covering automated taxi system in Zurich. *Transportation Research Part C: Emerging Technologies*, Vol. 123, p. 102974. https://doi.org/10.1016/j.trc.2021.102974

Hu, J., Saleem, A., You, S., Nordström, L., Lind, M., and Østergaard, J. (2015) A multi-agent system for distribution grid congestion management with electric vehicles. *Engineering Applications of Artificial Intelligence*, Vol. 38, pp. 45–58, https://doi.org/10.1016/j.engappai.2014.10.017

IDEO. (2015) *A field guide to human-centred design*. 1st ed.

Interaction Design (2020) User Centered Design https://www.interaction-design.org/literature/topics/user-centered-design [accessed November 2023]

Jacobs, J. (1961) *The death and life of Great American cities*. New York (NY): Random House.

Joubert, J. W. and de Waal, A. (2020) Activity-based travel demand generation using Bayesian networks. *Transportation Research Part C: Emerging Technologies*, Vol. 120, p. 102804. https://doi.org/10.1016/j.trc.2020.102804

Kaddoura, I., Bischoff, J., and Nagel, K. (2020) Towards welfare optimal operation of innovative mobility concepts: External cost pricing in a world of shared autonomous vehicles. *Transportation Research Part A: Policy and Practice*, Vol. 136, pp. 48–63. https://doi.org/10.1016/j.tra.2020.03.032

Korolko, N., and Sahinoglu, Z. (2015) Robust optimization of EV charging schedules in unregulated electricity markets. *IEEE Transactions on Smart Grid*, Vol. 8/1, pp. 149–157. https://doi.org/10.1109/TSG.2015.2472597

Kropotkin, P. (1902) *Mutual aid: A factor of evolution*. London: Freedom Press.

Lajas, R., and Macário, R. (2020). Public policy framework supporting "mobility-as-a-service" implementation. *Research in Transportation Economics*, Vol. 83, p. 100905. https://doi.org/10.1016/j.retrec.2020.100905.

Lascoumes, P. and Le Galès, P. (2005). *Gouverner par les instruments* (Vol. 200). Paris: Presses de Sciences Po.

Le Bescond, V., Can, A., Aumond, P., and Gastineau, P. (2021) Open-source modeling chain for the dynamic assessment of road traffic noise exposure. *Transportation Research Part D: Transport and Environment*, Vol. 94, p. 102793. https://doi.org/10.1016/j.trd.2021.102793.

Leblond, V., Desbureaux, L., and Bielecki, V. (2020) A new agent-based software for designing and optimizing emerging mobility services: application to city of Rennes. In European Transport Conference 2020.

Leng, N. and Corman, F. (2020) The role of information availability to passengers in public transport disruptions: An agent-based simulation approach. *Transportation Research Part A: Policy and Practice*, Vol. 133, pp. 214–236. https://doi.org/10.1016/j.tra.2020.01.007

Li, H., Wan, Z., and He, H. (2019) Constrained EV charging scheduling based on safe deep reinforcement learning. *IEEE Transactions on Smart Grid*, Vol. 11/3, pp. 2427–2439. https://doi.org/10.1109/TSG.2019.2955437

Lopez, P.A., Wiessner, E., Behrisch, M., Bieker-Walz, L., Erdmann, J., Flotterod, Y.-P., Hilbrich, R., Lucken, L., Rummel, J., and Wagner, P. (2018) Microscopic traffic simulation using SUMO, in: 2018 21st International Conference on Intelligent Transportation Systems (ITSC). Presented at the 2018 21st International Conference on Intelligent Transportation Systems (ITSC), IEEE, Maui, HI, pp. 2575–2582. https://doi.org/10.1109/ITSC.2018.8569938

Maciejewski, M., Bischoff, J., Hörl, S., and Nagel, K. (2017) Towards a testbed for dynamic vehicle routing algorithms. In: Bajo, J., Vale, Z., Hallenborg, K., Rocha, A.P., Mathieu, P., Pawlewski, P., Del Val, E., Novais, P., Lopes, F., Duque Méndez, N.D., Julián, V., and Holmgren, J. (eds.), *Highlights of practical applications of cyber-physical multi-agent systems, communications in computer and information science*. Cham: Springer International Publishing. https://doi.org/10.1007/978-3-319-60285-1_6

Maigha and Crow, M.L. (2014) Economic Scheduling of Residential Plug-In (Hybrid) Electric Vehicle (PHEV) Charging. *Energies*, Vol. 7, pp. 1876–1898. https://doi.org/10.3390/en7041876

Manser, P., Becker, H., Hörl, S., and Axhausen, K. W. (2020) Designing a large-scale public transport network using agent-based microsimulation. *Transportation Research Part A: Policy and Practice*, Vol 137, pp. 1–15, https://doi.org/10.1016/j.tra.2020.04.011.

Miaskiewicz, T. and Kozar, K. A. (2011) Personas and user-centered design: How can personas benefit product design processes? *Design Studies*, Vol. 32/5, pp. 417–430. https://doi.org/10.1016/j.destud.2011.03.003

Miskolczi, M., Földes, D., Munkácsy, A., and Jászberényi, M. (2021) Urban mobility scenarios until the 2030s. *Sustainable Cities and Society*, Vol. 72, p. 103029. https://doi.org/10.1016/j.scs.2021.103029.

Müller, K. (2017) *A generalized approach to population synthesis*. ETH Zurich. https://doi.org/10.3929/ETHZ-B-000171586

Namazi-Rad, M.-R., Tanton, R., Steel, D., Mokhtarian, P., and Das, S. (2017) An unconstrained statistical matching algorithm for combining individual and household level geo-specific census and survey data. *Computers, Environment and Urban Systems*, Vol. 63, pp. 3–14. https://doi.org/10.1016/j.compenvurbsys.2016.11.003

Newsham, G.R., and Bowker, B. G. (2010) The effect of utility time-varying pricing and load control strategies on residential summer peak electricity use: A review. *Energy Policy*, Vol. 38/7, pp. 3289–3296. https://doi.org/10.1016/j.enpol.2010.01.027

Pangbourne, K., Stead, D., Mladenović, M., and Milakis, D. (2018) The case of mobility as a service: A critical reflection on challenges for urban transport and mobility governance. *Governance of the Smart Mobility Transition*, pp. 33–48.

Piattoni, S. (2010) *The theory of multi-level governance: Conceptual, empirical, and normative challenges*. Oxford: Oxford University Press. https://doi.org/10.1093/acprof:oso/9780199562923.001.0001

Pruitt, J., and Grudin, J. (2003) Personas: Practice and theory, conference on designing for user experiences, San Francisco, June 2003, Association for Computing Machinery, New York (NY), pp. 1–15. https://doi.org/10.1145/997078.997089

Qian, T., Shao, C., Li, X., Wang, X., Chen, Z., and Shahidehpour, M. (2022) Multi-Agent Deep Reinforcement Learning Method for EV Charging Station Game. *IEEE Transactions on Power Systems*, Vol. 37/3, pp. 1682–1694. https://doi.org/10.1109/TPWRS.2021.3111014

Reckwitz, A. (2002) Toward a theory of social practices: A development in culturalist theorizing. *European Journal of Social Theory*, Vol. 5/2, pp. 243–263. https://doi.org/10.1177/13684310222225432

Rifkin, J. (2011) *The third industrial revolution*. London: Palgrave Macmillan.

Rohr, C., Ecola, L., Zmud, J., Dunkerly, F., Black, J., and Baker, E. (2016) *Travel in Britain in 2035: Future scenarios and their implications for technology innovation*. Santa Monica/Cambridge: RAND Corporation.

Roy, W., and Yvrande-Billon, A. (2007). Ownership, contractual practices and technical efficiency: The case of urban public transport in France. *Journal of Transport Economics and Policy*, Vol. 41/2, pp. 257–282.

Ruelens, F., et al. (2012) Demand side management of electric vehicles with uncertainty on arrival and departure times. 2012 3rd IEEE PES Innovative Smart Grid Technologies Europe (ISGT Europe), Berlin, Germany, pp. 1–8. https://doi.org/10.1109/ISGTEurope.2012.6465695.

Saadi, I., Mustafa, A., Teller, J., Farooq, B., and Cools, M. (2016) Hidden Markov Model-based population synthesis. *Transportation Research Part B: Methodological*, Vol. 90, pp. 1–21. https://doi.org/10.1016/j.trb.2016.04.007

Salet et al. (2003) *Metropolitan governance and spatial planning. Comparative case studies of European city-regions*. Spon-Press. Taylor and Francis group.

Sallard, A., Balać, M., and Hörl, S. (2021) An open data-driven approach for travel demand synthesis: An application to São Paulo. *Regional Studies, Regional Science*, Vol. 8, pp. 371–386. https://doi.org/10.1080/21681376.2021.1968941

Salminen, J., Santos, J.M., Jung, S.-G., Eslami, M., and Jansen, B.J. (2020) Persona transparency: Analyzing the impact of explanations on perceptions of data-driven personas. *International Journal of Human-Computer Interaction*, Vol. 36/8, pp. 788–800. https://doi.org/10.1080/10447318.2019.1688946

Sanders, E.B.-N., and Stappers, P.J. (2008) Co-creation and the new landscapes of design. *International Journal of Co Creation in Design and the Arts*, Vol. 4, pp. 5–18. https://doi.org/10.1080/15710880701875068

Sanders, E. B.-N., and Stappers, P. J. (2014) Probes, toolkits and prototypes: Three approaches to making in codesigning. *CoDesign*, Vol. 10/1, pp. 5–14, https://doi.org/10.1080/15710882.2014.888183

Schäfer, K., Rasche, P., Bröhl, C., Theis, S., Barton, L., Brandl, C., Wille, M, Nitsch, V., and Mertens, A. (2019) Survey-based personas for a target-group-specific consideration of elderly end users of information and communication systems in the German health-care sector. *International Journal of Medical Informatics*, Vol. 132, p. 103924. https://doi.org/10.1016/j.ijmedinf.2019.07.003

Seshadri, P., Joslyn, C., Hynes, M., and Reid, T. (2019) Compassionate design: Considerations that impact the users' dignity, empowerment and sense of security. *Design Science*, Vol. 5/E21. https://doi.org/10.1017/dsj.2019.18

Seyfang, G., and Smith, A. (2007) Grassroots innovations for sustainable development: Towards a new research and policy agenda. *Environmental Politics*, Vol. 16/4, pp. 584–603, https://doi.org/10.1080/09644010701419121.

Smith, G. (2020) Making mobility-as-a-service: Towards governance principles and pathways. Doctoral Dissertation. Chalmers University of Technology.

Soares, J., et al. (2017) Dynamic electricity pricing for electric vehicles using stochastic programming, Vol. 122, pp. 111–127.

Spaniol, M.J., and Rowland, N.J. (2018) Defining scenario. *Futures & Foresight Science*, Vol 1/1. https://doi.org/10.1002/ffo2.3

Stevenson, P.D., and Mattson, C.A. (2019) The personification of big data. International Conference on Engineering Design ICED19, Delft, August 2019. https://doi.org/10.1017/dsi.2019.409

Sun, L. and Erath, A. (2015) A Bayesian network approach for population synthesis. *Transportation Research Part C: Emerging Technologies*, Vol. 61, pp. 49–62. https://doi.org/10.1016/j.trc.2015.10.010

Suyono, H., Rahman, M.T., Mokhlis, H., Othman, M., Illias, H.A., and Mohamad, H. (2019) Optimal scheduling of plug-in electric vehicle charging including time-of-use tariff to minimize cost and system stress. *Energies*, Vol. 12/8. https://doi.org/10.3390/en12081500

Train, K. (2009) *Discrete choice methods with simulation*, 2th ed. Cambridge/New York: Cambridge University Press.

Tuchnitz, F., Ebell, N., Schlund, J., and Pruckner, M. (2021) Development and evaluation of a smart charging strategy for an electric vehicle fleet based on reinforcement learning. *Applied Energy*, Vol. 285, p. 116382. https://doi.org/10.1016/j.apenergy.2020.116382.

University of Cambridge. (2023) *Inclusive design toolkit*. University of Cambridge. http://www.inclusivedesigntoolkit.com/ [accessed August 2023]

Urry, J. (2016) *What is the future?* Cambridge: Policy Press.

Usman, M., Tareen, W.U.K., Amin, A., Ali, H., Bari, I., Sajid, M., Seyedmahmoudian, M., Stojcevski, A., Mahmood, A., and Mekhilef, S. (2021) A coordinated charging scheduling of electric vehicles considering optimal charging time for network power loss minimization. *Energies*, Vol. 14/17. https://doi.org/10.3390/en14175336

Vallet, F., Puchinger, J., Millonig, Al., and Lamé, G. (2020) Tangible futures: Combining scenario thinking and personas—A pilot study on urban mobility. *Futures*, Vol. 117/102513, pp. 1–26. https://doi.org/10.1016/j.futures.2020.102513

Vandael, S., Claessens, B., Ernst, D., Holvoet, T., and Deconinck, G. (2015) Reinforcement learning of heuristic EV fleet charging in a day-ahead electricity market. *IEEE Transactions on Smart Grid*, Vol. 6/4, pp. 1795–1805. https://doi.org/10.1109/TSG.2015.2393059

Verloo, N. (2019) Captured by bureaucracy: Street-level professionals mediating past, present and future knowledge, In: Raco, M. and Savini, F. (eds), *Planning and knowledge: How new forms of technocracy are shaping contemporary cities*. Policy Press, pp. 75–89.

Wan, Z., Li, J., He, H., and Prokhorov, D. (2018) Model-free real-time EV charging scheduling based on deep reinforcement learning. *IEEE Transactions on Smart Grid*, Vol. 10/5, pp. 5246–5257. https://doi.org/10.1109/TSG.2018.2879572

Wang, Z., and F. Li. (2011) Critical peak pricing tariff design for mass consumers in Great Britain. 2011 IEEE Power and Energy Society General Meeting.

Wang, R., Wang, P., Xiao, G., and Gong, S. (2014) Power demand and supply management in microgrids with uncertainties of renewable energies. *International Journal of Electrical Power & Energy Systems*, Vol. 63. pp. 260–269. https://doi.org/10.1016/j.ijepes.2014.05.067

Wang, R., Xiao, G., and Wang, P. (2017) Hybrid centralized-decentralized (HCD) charging control of electric vehicles. *IEEE Transactions on Vehicular Technology*, Vol. 66/8, pp. 6728–6741.

Wang, F., Gao, J., Li, M., and Zhao, L. (2020) Autonomous PEV charging scheduling using Dyna-Q reinforcement learning. *IEEE Transactions on Vehicular Technology*, Vol. 69/11, pp. 12609–12620. https://doi.org/10.1109/TVT.2020.3026004.

Watson, V. (2002) Do we learn from planning practice? The contribution of the "practice movement" to planning theory. *Journal of Planning Education and Research*, Vol. 22/2, pp. 178–187. https://doi.org/10.1177/0739456X02238446

Watson, V. (2003) Conflicting rationalities: Implications for planning theory and ethics. *Planning Theory & Practice*, Vol. 4/4, pp. 395–407. https://doi.org/10.1080/1464935032000146318

Watson, M. (2012) How theories of practice can inform transition to a decarbonised transport system. *Journal of Transport Geography*, Vol. 24, pp. 488–496. https://doi.org/10.1016/j.jtrangeo.2012.04.002

Wildfire, C. (2018) How can we spearhead city-scale digital twins? *Infrastructure Intelligence*. http://www.infrastructure-intelligence.com/article/may-2018/how-can-we-spearhead-city-scale-digital-twins [accessed August 2023]

Yameogo, B.F., Vandanjon, P.-O., Gastineau, P., and Hankach, P. (2021) Generating a two-layered synthetic population for French municipalities: Results and evaluation of four synthetic reconstruction methods. *Journal of Artificial Societies and Social Simulation*, Vol. 24/5. https://doi.org/10.18564/jasss.4482

Yang, Y., Jia, Q.-S., Deconinck, G., Guan, X., Qiu, Z., and Hu, Z. (2019) Distributed Coordination of EV Charging with Renewable Energy in a Microgrid of Buildings. IEEE Power & Energy Society General Meeting (PESGM), Atlanta, pp. 1–1. https://doi.org/10.1109/PESGM40551.2019.8973991

Yin, Y., Zhou, M., and Li, G. (2015) Dynamic decision model of critical peak pricing considering electric vehicles' charging load. International Conference on Renewable Power Generation (RPG 2015), Beijing, pp. 1–6. https://doi.org/10.1049/cp.2015.0564

Zhang, X., Liang, Y., and Liu, W. (2017) Pricing model for the charging of electric vehicles based on system dynamics in Beijing. *Energy*, Vol. 119, pp. 218–234. https://doi.org/10.1016/j.energy.2016.12.057

Ziemke, D., Charlton, B., Hörl, S., and Nagel, K. (2021) An efficient approach to create agent-based transport simulation scenarios based on ubiquitous Big Data and a new, aspatial activity-scheduling model. *Transportation Research Procedia*, Vol. 52, pp. 613–620. https://doi.org/10.1016/j.trpro.2021.01.073

CHAPTER 4

A Holistic Sustainable Transition Approach: Theory to Action

Abstract This chapter brings together the trends and methods and presents three case studies where challenges of urban mobility are addressed through mixed methods approaches. The first case study assesses the socio-environmental impacts of shared automated electric vehicles in Paris. The second looks at ways how active mobility—and walking in particular—could be better integrated in Mobility-as-a-Service solutions. Finally, we compare three promising interventions to improve the environmental and social performance of the urban mobility system of Cairo. Together, the three projects showcase some of the potentials of transdisciplinary approaches and how they can be used to tackle complex challenges with multiple, partially competing objectives.

Keywords Shared vehicles · Automated vehicles · Mobility-as-a-Service · Bus Rapid Transit · Walkability · Intermodality · Paris · Cairo

In this chapter, we look ahead and present practical examples of more integrated and transdisciplinary approaches. The overarching objective remains to create more sustainable and people-centred urban mobility systems for tomorrow. Referring to the introduced trends and uncertainties, we are making use of a result-based approach and apply it to core

T. Gall et al., *Sustainable Urban Mobility Futures*, Sustainable Urban Futures, https://doi.org/10.1007/978-3-031-45795-1_4

urban mobility challenges in specific contexts. A result-based management approach (RBM) prioritises achieving predefined results compared to an often-dominating focus on resources and activities (Babst et al., 2022). This approach involves setting clear objectives, tracking progress towards those objectives, and adjusting strategies as needed to ensure that the desired results are achieved. We make use of this approach for two reasons: First, resources in an RBM are allocated to maximise their impact. Further, such approaches are usually more adaptable than traditional approaches as they do not require sticking to rigid processes, specific disciplines, or methods. In our context, we want to create positive impacts on specific transitions within complex urban mobility systems. For this, the RBM helps to focus on the goal that shall be achieved and then choosing a combination of adequate conceptual framings, disciplines, and methods to get there. Comparing to academic epistemologies, a pragmatic approach shows many similarities and can be referred to for further discussion.

The described cases have been developed as part of the Anthropolis Chair[1] which works on approaches to support the design of people-centred mobility solutions whilst addressing three interconnected areas of interest and a transversal theme (Fig. 4.1). The first focus area of the Chair is that of *urban life and mobility futures*. Within this scope, we work on supporting the process of generating localised future scenarios whilst creating knowledge on future mobility behaviours and practices. The second subject relates to the development of MaaS, and more specifically on the conditions that could lead to sustainable value creation for a mobility ecosystem. The third axis concentrates on future mobility infrastructures, for instance, related to public transport projects in Greater Paris. The transversal theme looks at sustainability across the topics. Practically, this was translated into: (1) environmental and health indicators added to the development of sets of 2030 future scenarios; (2) The investigation of sustainable business models for MaaS; (3) The elaboration of guidelines to ensure the sustainability of mobility infrastructures. The scope of the research work entails that of people's daily mobility in the geographic areas of Greater Paris and Greater Cairo.

[1] Research chair managed by the Technological Research Institute IRT SystemX and the Industrial Engineering Laboratory at CentraleSupélec, Université Paris-Saclay, with the partners Groupe Renault, Nokia Bell Labs, EDF, Engie, and the inter-council partnership Paris-Saclay.

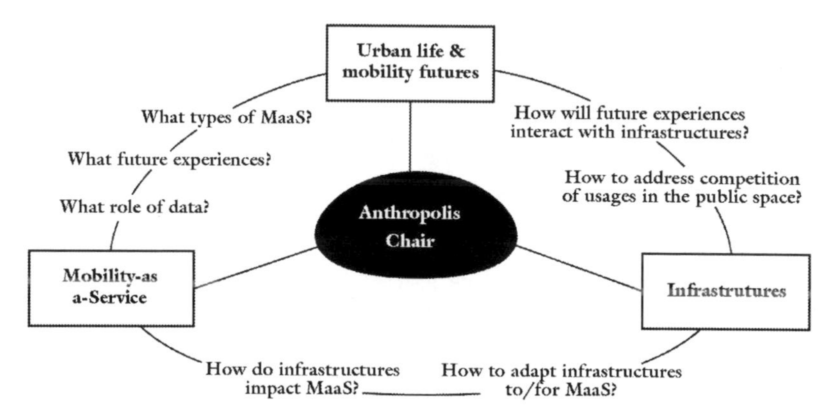

Fig. 4.1 Connections of topics addressed at the Anthropolis Chair

Using a simplified RBM framework, we start with the identification of concerned stakeholders and problems to be addressed. In the next step, we select how we can define the objective or the anticipated direction, as well as a way to measure the progress. Next, we choose the inputs, methods, and approaches that are most adequate to achieve the stated objective(s). Finally, after the implementation or execution of the developed approach, an evaluation and learning stage follows which can provide input into another RBM cycle. This possibly sounds logical on the one hand but, oftentimes, is not. Instead, fixed procedures, bureaucratic processes, or disciplinary blinders prevail. On the other hand, it might appear very conceptual as described above. We aim to highlight its key elements, stages, and advantages through three specific examples in the following part.

The problems we discuss here aim to span a range of possible applications, show different socio-cultural contexts and scales, as well as being representative for areas that require transdisciplinary and multi-method responses (Table 4.1). The first is the challenge to evaluate how different variations of Shared Automated Electric Vehicles (SAEVs) can respond to different mobility needs in a peri-urban context just south of the capital of France—the inter-council partnership Paris-Saclay. This problem builds on various works conducted as part of the Anthropolis Chair.

We define the second problem as the lack of active mobility, in particular, walking and personal bike use, as part of new, primarily digital

Table 4.1 Examples of applications addressing different trends and uncertainties

Application cases		*1. SAEV in peri-urban Paris*	*2. Active mobility in MaaS*	*3. Interventions in Cairo*
Methods		Persona-based design Agent-based simulation Comparison of future scenarios	User experience Design Stakeholder analysis Agent-based simulation	Persona-based design Agent-based simulation Comparison of future scenarios
Societal trends	Population growth, ageing	X		X
	Behaviours		X	
	Hypermobility	X		
	Sedentary lifestyles		X	
Urban trends	Urbanisation		X	X
	Glocalisation	X		
	Agglomeration economies	X	X	
	Sprawl vs compactness		X	X
	Transit-oriented development	X		X
Technological trends	Alternative energy sources	X		
	Automated vehicles	X		
	Technology in cities		X	X
	Platformisation		X	X
	ITS		X	X

solutions for enabling multimodal trips. The challenge lies at the fact that walking and personal bike use, in contrast to other mobility modes, do not provide immediate economic return for any stakeholder. A focus lies thus on enablers for a holistic, sustainable, and people-centred concept of MaaS—discussed on the European scale of urban areas with zoom-ins on Sweden and France.

Finally, we tackle the challenge of an urban mobility system transition in the metropolitan area of Cairo. The specific problem is to decide, within a very uncertain and complex context, which types and characteristics of solutions can have the highest impact towards mass transit use within limited resources. We look at the impact of Bus Rapid Transit (BRT), urban design for active mobility around stations, and station design and management to enable intermodality.

For each of the three problems, we discuss the problem, concerned stakeholders (public, private, academia, and civil society), and suitable methods. Whilst referring to more detailed discussions, we outline the applied mixed-method approaches, combining a response to trends and uncertainties introduced in Chapter 2, and using a combination of methodologies presented in Chapter 3.

4.1 Shared Automated Electric Vehicles for Better Accessibility Around Paris

Our first practical example is the assessment of potentials of different variations of SAEVs. As discussed in the earlier sections, electric and autonomous vehicles are a central theme in technological innovation. On the one hand, EVs can reduce significantly GHG emissions per km. On the other hand, AVs can reduce costs for smaller shuttles and on-demand services, making increased accessibility primarily in rural and peri-urban areas more possible. Various case studies and innovation projects (Vosooghi et al., 2019, 2020; Horschutz Nemoto et al., 2021) test the suitability of SAEV with varying capacity (usually between 4 and 12) to connect to existing sustainable transport infrastructure, such as rail-based mass transport. Despite remaining technical challenges regarding AVs market-readiness and discussions around EV rare material constraints and geopolitical dependencies, as well as local pollution through tire abrasion, a significant potential is seen in their large-scale uptake.

4.1.1 Problem Definition

Whilst this potential is there, many questions can be posed. We propose a transdisciplinary method to answer some of them. First, it is a question of location: Where are the vehicles positioned and how far can they go? Next, there is the fleet size and constellation: How many vehicles are needed in a certain area to achieve the aimed for service level? Such a service level can be measured, for example, by the percentage of requested trips that must be rejected due to vehicle unavailability or by high-waiting times. Third, ongoing pilots and planned projects often consider a public private partnership and public subsidies, for example, to serve areas that can currently not be reached with existing public transport. To evaluate the usefulness and extent of the contribution of SAEVs in specific areas to defined public objectives, we require an approach that can measure this potential impact, across environmental, social, and economic dimensions. Such evaluation can only be done in a future setting, requiring the consideration of trends and uncertainties, such as *growing and ageing population*, and varying levels of, e.g., *private car ownership*. These three questions only constitute a small section of the real complexity, yet we claim that finding some answers to them can already significantly contribute to decision-making and possible SAEV service design at a local level.

4.1.2 Stakeholders

A set of stakeholders are concerned by the implementation of SAEVs. This includes foremost the user, as well as people potentially affected by it (e.g., via changing traffic congestion/risk), the private sector as service and technology provider, the public sector across scales as potential client and regulating body, and existing public transport providers as potential competitors and collaborators in case of intermodal mobility. We focus here on the individuals affected by it for two reasons. First, the goal shall primarily benefit individuals within environmental constraints. Thus, being able to measure this, is of importance for all stakeholders. Second, other stakeholders have diverging and sometimes not known objectives which make a general assessment of impacts challenging. In the following, the focus thus lies on the different individuals whilst considering their interdependence with other stakeholders.

4.1.3 Impacts and Assessment

Assessing current and potential future situations to compare them and make data-driven decisions is a practice that has defined transport and mobility research for a long time. From the beginning, the 'throughput' was mostly prioritised, meaning how many vehicles can be moved through a certain link. This might be the number of passengers in a metro system or the cars on a road. However, in the more recent past, assessment diversified. First, environmental sustainability discussions around

GHG emissions, local pollution, and noise have entered the equations. Furthermore, an increasing focus has been on the individual user and their choices. Initially on the time and costs, later extended with factors such as comfort and safety, in particular, regarding a stronger focus on other groups than the initial primary targeted user—working men going to work. Lastly, measures such as more complex accessibility to job markets, health and education facilities, green and public space, amongst others, were integrated.

To operationalise this, various frameworks have been developed. The leading ones are the triple-bottom line of sustainability with the environmental, social, and economic dimension, or its stylised private-sector equivalent People-Planet-Profit. In the mobility sector, two recent compilations of indicators by Chatziioannou et al. (2023) and Xenou et al. (2021) are noteworthy, as well as a targeted assessment framework for SAEVs (Horschutz-Nemoto et al., 2021). Each of them contains a set of indicators with various overlaps. The most commonly referenced indicators are those of environmental sustainability expressed via potential *GHG emissions*, social sustainability via *accessibility to opportunity*, and economic sustainability via the *mobility-related expenses*. The latter are often extended by costs of the maintenance and construction of infrastructure. However, our focus here lies on the individual choices' impacts, thus we exclude them from the calculations but suggest their consideration afterwards in the design and decision-making process.

Box 4.1 CO_2 emissions per mode and CO_2 budgets per person [facts]

About 1.5–2 tons CO_2e per person and year are often quoted as maximum value if we want to remain under an increased temperature of +1.5 °C compared to preindustrial values. Today's per capita emissions range between 4.5 tons for bottom-up models and 12 or more for top-down models. The largest contributors are usually housing, mobility, and nutrition. The graphic below shows the CO_2e emissions per passenger per kilometre for some of the most common modes.

*

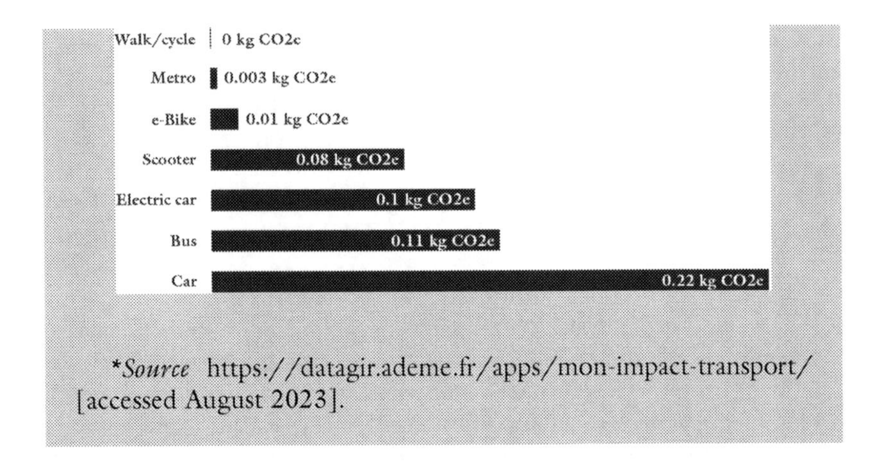

**Source* https://datagir.ademe.fr/apps/mon-impact-transport/ [accessed August 2023].

We focus here on three indicators to assess the impact at the Paris-Saclay scale for the present and four 2030 scenarios with and without SAEV. The first indicator is on *emissions*, commonly referred to as Greenhouse Gas emissions or CO_2 equivalent (CO_2e) emissions. In both cases, different emissions—primarily CO_2, methane, and nitrous oxide in the transport context—are transformed to an assumed equivalent of CO_2 to permit easier impact comparisons. As our simulations run for 24 hours, we provide daily averages or sums, expressed as CO_2e from personal mobility between 0:00 and 23:59 at an average weekday. The 2030 scenarios are calculated individually as well as an average of the four scenarios with equal weight (expressing equal probability). The emissions are calculated on mode-dependent CO_2e-averages in kilograms per passenger kilometre (pkm). For example, a bus with average emissions of 1 kg CO_2e per kilometre and an average occupancy of 50 people has a value of 0.02 kg CO_2e/pkm. The values are resulting from current fleet constellations and their projections until 2030. Box 4.1 shows an overview of various modes and their equivalent CO_2e values for France.

The second indicator is *accessibility* as the direct impact for the people moving around. An urban area functions as a spatial container for individuals to collaborate more and use resources more efficiently and effectively due to physical proximity. In the past (before 1860s), this physical proximity was defined by the duration which could be walked per day without negatively impacting the time needed to work, sleep, eat, etc. The duration per day is usually defined somewhere around 60, maximum

90 minutes per day, depending on the scale of city and with large global homogeneity despite minor cultural differences (see Marchetti, 1994, for initial analysis of globally homogeneous commuting times, Bertraud, 2018, for urban growth dynamics and commuting durations, and Dong et al., 2022, for recent, data-driven analysis of daily commuting patterns). We thus aim to measure accessibility as an indicator aggregated over the population for reaching jobs and other opportunities in the urban context. Geurs and van Wee define accessibility as the 'extent to which land-use and transport systems enable (groups of) individuals to reach activities or destinations by means of a (combination of) transport mode(s)' (Geurs and van Wee, 2004, p. 128).

As our focus on the impact of SAEVs, the question is to measure the change of accessibility across scenarios and with or without the service. Most commonly, accessibility is measured as potential reach of geographical area or number of jobs/places. As outcomes of the simulations, we obtain actual numbers of specific trip times per individual. Thus, we need to adapt and calculate a proxy value which we set at how long it takes to get somewhere and how far the actual distance between the locations is, taking into consideration that locations do not change due to SAEVs, only across scenarios as result of different synthetic populations. A limitation of this approach is, amongst others, pointed out by Bertraud (2018, Ch. 5), who states that 'an urban transport system that would solely minimise travel time between home and current jobs for all workers would result in poor mobility, as in the future, workers might not be able to reach many alternative jobs that would improve their job satisfaction or salary'. Whilst being aware of the limitation, we argue that the aggregated or averages values of individuals whose trip origins and destinations are distributed across the geographical area with representative densities act as sufficiently close proxy accessibility scale.

The distances and durations, disaggregated by mode, are calculated per day. All values and settings are identical to those described for CO_2e emissions. The following part shows the respective formulas for the assessment.

Finally, the actual *individual costs* are looked at. The values are given in Euros and include direct costs (i.e., metro tickets, fuel costs, monetised car wear and tear) but do not consider generalised expenses such as road construction and maintenance or public transport expansion. We assume walking, biking, and using public transport for those having a public transport subscription as free. Car trips are calculated at 90 cents per kilometre, passenger trips at half the value. Public transport is set at €2 per trip, capped at a daily maximum of €4 as otherwise a monthly subscription would be cheaper. SAEVs prices are building on the system of current taxi pricing scheme in the Paris region.[2] We do not consider a minimum trip cost to enable short distance trips between homes and public transport hubs. As SAEVs are supposed to be cheaper than traditional taxis due to sharing between customers, more efficient engines/ driving styles resulting from automated mobility, and potential public subsidy due to suburban setting, we assume a quarter of the current taxi prices, thus 0.65 cents as a pickup fee and 0.38 cents per kilometre.

4.1.4 Methodology

We make use of a mixed approach to simulate the urban mobility system of 2030 for the Île-de-France region using MATSim, a multi-agent transport simulation toolkit. The analysis considers four base scenarios incorporating the new metro line 18 and tram 12 which are part of the Grand Paris Express whilst integrating additional trends and uncertainties through four different synthetic populations (Gall et al., 2023). Additionally, a second set of four simulations is performed by introducing an on-demand mobility solution with SAEVs.

The Île-de-France region, and the fast-developing area of Paris-Saclay (Fig. 4.2), are facing significant urban mobility challenges due to rapid population growth, congestion, and environmental concerns. To address these issues, the region is investing in public transport infrastructure, such as the metro line 18, the south most extension of the Grand Paris Express. In addition, emerging mobility solutions, such as SAEVs are considered by public and private actors to play an essential role in the future urban mobility ecosystem. A particular potential benefit is seen in the intermodal connection between areas that have been traditionally and still today remain to be highly dependent on personal vehicles and stations of mass transit (RER/metro/tram).

To quantify the potential impacts—including its direct impacts such as reduced costs or increased accessibility or indirect ones such as GHG emissions—and thus assess the effectiveness of these interventions, this

[2] €1.53 per km for zone b in Paris region according to G7 data on 18 May 2023 (https://www.g7.fr/en/paris-taxi-fares) and pick-up fee of €2.6.

Fig. 4.2 Map of Paris-Saclay (bold outline) south of Paris, showing primary road infrastructure (grey), and rail-based transport infrastructure in 2030 (*Source* OpenStreetMap, 2022)

study employs MATSim, an open-source, multi-agent transport simulation toolkit, to simulate the urban mobility system of the Île-de-France region for 2030. It models urban mobility based on individual agents, thus people living and moving in a particular area, during a full standard weekday (00:00–23:59), taking into consideration the road and public transport network, its GTFS schedule, locations of work, education, commerce, etc., as well as an underlying choice model. Calibrated via actual numbers or aggregate values, this makes up a set of input values that are fed into the model. This is run n iterations (in our case 100) to find an equilibrium for the scoring factors (e.g., reduction of time and expenses). After each iteration, up to 5% of the simulated population can change their choice to improve their individual utility.

Due to the high-computational requirements to simulate 100 iterations of the daily movements in the Île-de-France region with a population of about 14 M, the *synthetic population* is oftentimes proportionally downscaled via random sampling. Whilst this reduces the accuracy of the results, it allows to run more complex and exploratory simulations but must be kept in mind. Further, it is important to note for which fields of application MATSim bears advantages. The fact that it permits modelling on the individual level invites any kind of applications that tries to understand different profiles and—importantly—their interaction via congestion or shared use of SAEVs. Thus, it fills a gap between higher level models such as most established four-step model which models aggregated demand, or higher detail models looking at intersections or any forms of mobility where mobility space is conceptualised more accurately as lines and nodes with defined capacities.

For the method described here, we use a two-step approach to simulate the urban mobility system of 2030. First, we create *four base scenarios* considering the new metro line 18 and tram 12 as well as trends and uncertainties by generating four different synthetic populations. These populations represent various demographic, socio-economic, and spatial distribution patterns based on existing data sources and projections for the year 2030. In the second step, we introduce an on-demand mobility solution using SAEVs and run an additional set of four simulations. This approach allows us to evaluate the potential impacts of SAEVs on travel behaviour, mode choice, congestion, and greenhouse gas emissions, as well as their synergies with public transport infrastructure. The simulation runs with 1% of the overall population, deemed sufficient for accurate comparisons but not necessarily to deduct the most accurate stand-alone values. Each SAEV has a passenger capacity of four people, and 200 shuttles are introduced in the Paris-Saclay area (equalling less than 20,000 vehicles for a 100% synthetic population due to higher ride-sharing rate). At the start of each simulation circle, the shuttles are located at one of 100 demand peak locations resulting from the base contextual scenarios. In between trips, shuttles are automatically rebalanced according to probable future trip origin locations. The SAEVs can only be used to go from or to rail-based public transportation, not to go directly from trip origin to destination.

4.1.5 Outcome and Recommendations

Restricted by the scope of work, we here only discuss two of the results. For this, we directly zoom in on the trips starting and/or ending in Paris-Saclay. Figure 4.3 indicates that currently, 697 tonnes CO_2e are emitted per day resulting from personal mobility trips starting and/or ending inside the inter-council partnership Paris-Saclay. The largest contributor

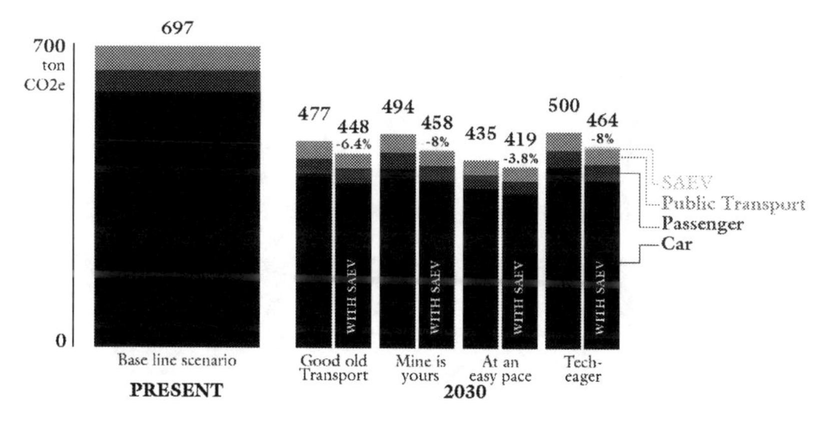

Fig. 4.3 Daily mobility emissions in Paris-Saclay. Impact of Shared Automated Electric Vehicles (SAEV), assuming CO_2e/pkm reduction by 2030

is the individual car, followed by car passengers and public transport. In the 2030 scenarios with SAEVs (rightmost bar), a small addition of emissions from SAEVs leads to a significant decrease of CO_2e emissions from individual cars. The total decrease of emissions due to SAEVs ranges from −3.8% in the 'At-an-easy-pace' scenario up to −8% for the 'Mine-is-yours' and 'Tech-eager' scenarios.

Aside from the data aggregated by scenario groups or scenarios, we can look at the impact of the individual personas (see Fig. 4.4 for first four personas). For each of the 16 personas, a fictional name and data-based information on age and occupation are provided. The occurrence of each persona today and averaged across the 2030 scenarios is presented below, restricted to those taking trips within CPS. On the right, the average values for all 2030 scenarios are provided for CO_2e emissions, average daily distance, time spent commuting, and expenses. The bars show the respective variations for each persona. The impact of SAEVs changes significantly across personas, affecting primarily those personas that are emitting more than average, thus those that are more car dependent.

As an example, we describe the meaning of the diagram for Persona 1 'Céline Dupont'. Currently, about 121,000 individuals of that type are moving in the area of Paris-Saclay on a normal day. By 2030, an across-scenario average of about 141,000, thus an increase of about 17%

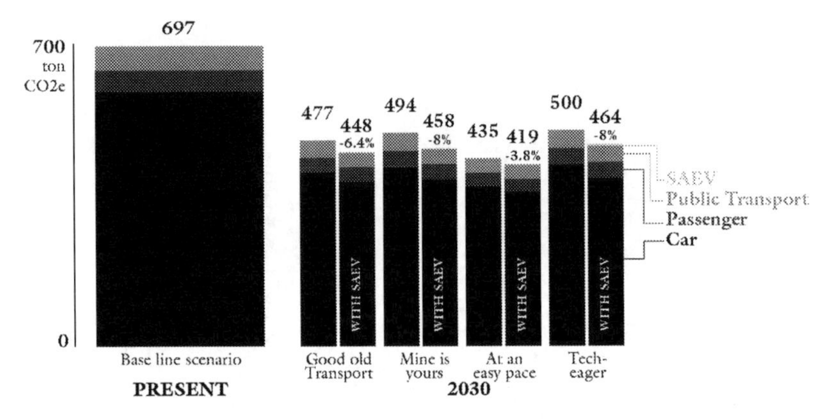

Fig. 4.4 Daily values for selection of four of 16 personas. Impact of Shared Automated Electric Vehicles (SAEV), assuming CO_2e/pkm reduction by 2030

is assumed. Compared to the daily 2030 average of 0.9 kg CO_2e emissions per day, Céline emits about 0.4 kg less. In 2030, regardless of SAEV or not, her daily CO_2e emissions decrease by 0.2–0.3 kg. A difference between the 2030 scenario sets can be observed in the other three values. For example, today, Céline spends about 19 minutes less than the 2030 average of 65 minutes per day for mobility. For the 2030 scenario without SAEV, this increases by 1 minute to 18 minutes less, or 47 minutes. In the 2030 scenarios with SAEV, the daily time savings increase to minus 22 minutes, or 43 minutes. This stems most likely from a potential combination of using SAEVs and/or indirect benefits from less traffic due to SAEVs and thus less cars. Similar observations can be made for the distance and costs.

In this application, we outlined a multi-step approach to integrate future uncertainty in the design and decision-making processes in complex urban mobility systems. We used methods to integrate trends and uncertainties from qualitative future scenarios into urban mobility system models by using personas and synthetic populations as intermediary design objects. Next, we applied this approach to agent-based mobility simulations at the scale of the Île-de-France region with a zoom-in on Paris-Saclay. Aside from the present simulation, four 2030 scenarios were simulated without and with SAEVs introduced in Paris-Saclay. Making use of a multi-dimensional impact assessment approach, we compared a few key indicators across scenario groups, scenarios, and persona: CO_2e emissions, time, and duration spent with mobility per day, and mobility-related expenses.

The results permit to quantify the potential impact of a new solution whilst taking future uncertainties into consideration. Three findings that can inform policy and design recommendations arise. First, the highest impact does not result from the tested solution but instead from the differences between scenarios, i.e., how many people of what persona live where. Secondly, the impact of energy efficiency of ICE vehicles is a critical value with significant impact across scenarios and solution variants. Finally, the difference between personas is significant across the tested dimensions. Future work exploring how to support the most disadvantaged people or redistribute costs and advantages using such persona-based approaches can contribute to the challenges of *mobility justice*.

Concluding, we argue that the presented methodology has the potential to integrate qualitative uncertainty in a structured and logical manner without making the process too resource demanding. It must be noted that the proposition is made at a relatively high level combining various approaches and disciplines. Thus, the main contribution is rather at a meta-methodological level than a detailed technical proposition in any of the sub-method (with exception on the persona and synthetic population transformation).

Some limitations remain. The focus of such numeric analysis should not be on the absolute numbers due to the high number of qualitative and quantified assumptions. Instead, the contribution can bring predominantly benefits in comparative situations. For example, comparing mobility service solutions against each other, across scenarios, or against not having them. Nevertheless, the complex system dynamics with multiple feedback loops and near-endless connections, external impacts, and so-called 'unknown unknowns' of the future make any prospective work prone to errors. Whilst this is acknowledged, we postulate that the adequate response is finding better tools and methods to integrate future uncertainty in today's practice and continuously evaluate and improve such methods when more or better information are available.

4.2 Active Mobility in Mobility-as-a-Service Solutions

The second case study stays in the same geographical context of Paris and takes up some of the already introduced and applied methods. However, the focus shifts to the role of active mobility in MaaS and how it can be strengthened. This sub-chapter follows the same RBM-inspired structure and contains several references to further works.

4.2.1 Problem Definition

MaaS is an innovation aiming to tackle issues of urban congestion and pollution through the facilitation of access to a larger panoply of mobility modes and to information. MaaS goals are linked to achieving SDGs by optimising mobility access. Nevertheless, this optimisation requires complex actions from all the actors in MaaS ecosystems. A key challenge for MaaS is the enhanced integration of active modes. Active modes like walking and cycling have the potential to help reduce short and medium trips made in polluting individual vehicles. The encouragement of walking and cycling would have positive impacts on health and the environment.

Cycling found a way into MaaS in France through shared bike services like Jump, Lime, or its dockless systems predecessors like GoBee Bike, Obike, Ofo, and Mobike between 2016 and 2017 (L'usine nouvelle, 2018). On the other hand, walking is not backed up by a service industry, and got left behind in that sense. One of the identified causes is the perceived lack of value creation potential that walking has in the eye of actors in MaaS ecosystems (see interviews in Reyes Madrigal, 2024). However, active modes account for many potentials if integrated to MaaS solutions. For example, boosting intermodal access to other modes of transport to allow longer distance trips and decrease the use of the private car, improving mobility justice regarding accessibility, permitting the provision of customised information, and the allocation of incentives. These spatial and welfare potentials are important pathways towards sustainability that could be enhanced through MaaS. However, the lack of consideration of pedestrian mobility blocks the overall purpose of MaaS of providing individuals an integrated 'seamless' access to mobility for their trips, encouraging more sustainable mobility practices. One of the main counterarguments to our hypothesis has been that 'walking is already present in MaaS platforms'. As a matter of fact, basic information for pedestrian itineraries is already provided in MaaS, nevertheless, four main aspects significantly lack attention.

The first is *how information is provided*. This refers to the universality of access to pedestrian-related information in itinerary suggestions and the overall MaaS platforms' User Interface (UI). Different types of platforms (private/public, local/generic) have different prioritisations of modes. Bonjour RATP, the app of the Parisian public transport operator, prioritises public transport, followed by walking, biking, e-kick scooter, and ride hailing. On the other hand, Google Maps puts private cars first, followed by motorcycles/scooters, public transport, walking, ride hailing, and biking (Reyes Madrigal et al., 2023).

The second is the *type and quality of the data* provided. This denotes the characteristics, accuracy, and usefulness of data obtainable (Fig. 4.5).

The third is the *possibility* to customise the user's profile to receive information according to its individual needs and capacities. For example, the IDFM App permits (state 01/2023) to choose between three walk speeds: 'Good walker', 'Normal walker', 'Walker with difficulties'.

The fourth is the *understanding and appraising walking as a tool* to generate and capture *sustainable value*. We define sustainable value as

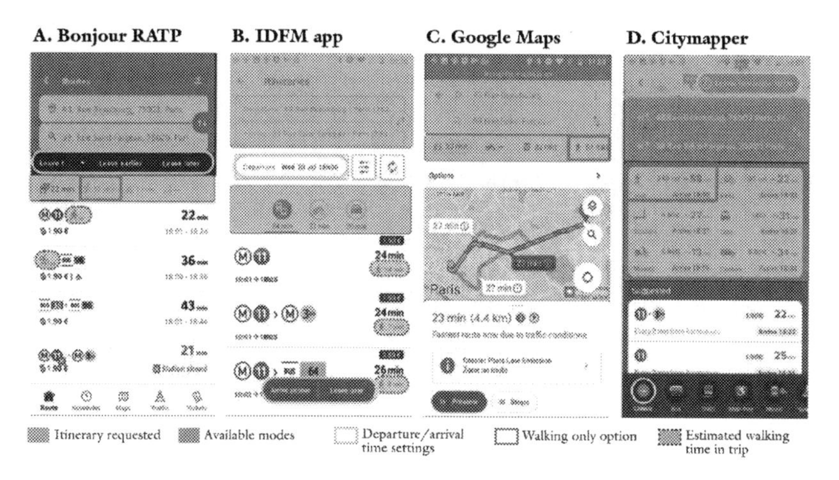

Fig. 4.5 Presence of walking-related information in the itinerary requests (state 01/2023)

the gains that may be considered indirect at first, but that bring significant benefits to society. These are usually related with health and the environment (e.g., physical and mental health, cleaner air).

> Pedestrians are unfamiliar figures, we do not know them well, we don't know how they move, where they are, where they go; and their different profiles and speeds. (expert interview, in Reyes Madrigal, 2024)

Currently, these four aspects remain underexplored and represent a research gap. Since they all rely on stakeholder decisions, we address these through the perspective of the stakeholders of the (enlarged) MaaS ecosystem of the Paris region. The need to territorialise the research problem came from the fact that MaaS exists only if attached to specific transport and service networks and individuals in societies willing to use them. Google Maps, an international solution, still requires local public transport information such as schedules and routes to function. To explore ways to address these challenges, we identify existing levers (power, interest) of MaaS stakeholders to better integrate walking in MaaS and tackle the outlined gaps.

4.2.2 *Stakeholders*

MaaS requires functional relationships amongst different actors and stakeholders in the mobility ecosystems of territories. These actors aim at creating value by working together to integrate different mobility services (and other added value services) and provide individuals with a new multimodal service; MaaS. We adopt the definition of ecosystems as *'communities of interdependent yet hierarchically independent heterogeneous participants who collectively generate an ecosystem value proposition [...and] often emerge through collective action, where ecosystem participants interact with each other and the external environment'* (Thomas and Ritala, 2022). The ecosystem of stakeholders of MaaS is composed of actors within the world of transport and mobility such as the PTA, MaaS operators, mobility service providers, but also outside of it, for example, actors in institutions of public health, commerce, finance, political sectors, academy, or consulting (Table 4.2). Here, we are interested in the decision-making dynamics between institutional stakeholders and the impacts on the choices of individuals to use MaaS and/or to walk (Fig. 4.6). This analysis confirms the lack of links between walking and

MaaS and led to the focus on potential levers amongst key stakeholders to better integrate walking in MaaS.

In the French context, the identified key stakeholders involved in the integration and development of active mobility in MaaS are local public authorities (cities, departments, metropolitan areas, regions), since they have the competency (responsibility) of maintaining and, in some cases, developing the roads' network, to which pedestrian infrastructure is traditionally attached. Additionally, national public authorities have a key role, as they have promoted new mobility guidelines through legislation and by doing so, they have paved the way to MaaS through ensuring accessibility in data provision and data sharing amongst the mobility and transport players. PTAs have a big role since they can mobilise resources to improve the four aspects: user experience, data available, customise profiles, and generate sustainable value from it all. The other stakeholders' roles and potential levers found can be found in Table 4.3.

4.2.3 Impacts and Assessment

This research addresses the topic of pedestrian mobility in MaaS by answering the research questions *how to improve the integration of walking in MaaS* and *how to measure the impacts of the integration of walking in MaaS*? These approaches, improvement, and measurement, require to be tackled with different perspectives. To answer the first question on the improvement of walking in MaaS, the identification of variables related to *digital accessibility* and *physical accessibility* is necessary.

The notion of *digital accessibility* comprises two layers. One refers to the functionalities that permit MaaS users to easily navigate in the MaaS platform and customise their trips, including the pedestrian parts of their trips. When discussing digital accessibility, we talk about user experience, user interfaces, universal design principles, and user customisation potentials. The second layer of digital accessibility deals with the availability of quality data of the territory where MaaS is implemented.

Physical accessibility, on the other hand, refers to the quality and availability of physical infrastructure, very much related to the concept of walkability. Knapskog et al. (2019) defined walkability as the 'extent the surroundings are nice to walk in, as well as pleasant and interesting, and inviting walking'. The authors consider criteria, such as 'infrastructure and traffic, urbanity, and surroundings and activities' to measure walkability.

Table 4.2 Key stakeholders' competencies to tackle the four central aspects of walking integration in MaaS (Reyes Madrigal, 2024)

Stakeholders' roles	*Levers*			
	How is information provided? [UX/UI]	*What information is provided? (Availability and quality of data)*	*To whom is the information provided? (Customisation potential)*	*Generating and capturing sustainable value through walking (why?)*
Transport authorities	X	X	X	X
Health authorities			X	X
National governments	X	X		X
Local/regional governments		X	X	X
European authority	X	X		
Transport regulation agency	X	X		X
NGOs		X	X	
Users/ individuals	X	X	X	X
MaaS operator	X		X	X
MaaS platform developer	X		X	X
Mobility service provider		X	X	
Public transport operator		X	X	X
Energy service provider				X
Chamber of commerce (enterprises)		X		X
Investors				X
Vehicle manufacturers				X
Parking operators		X		
Technology provider (payment, other)		X	X	
Researchers	X			
Consultants	X			

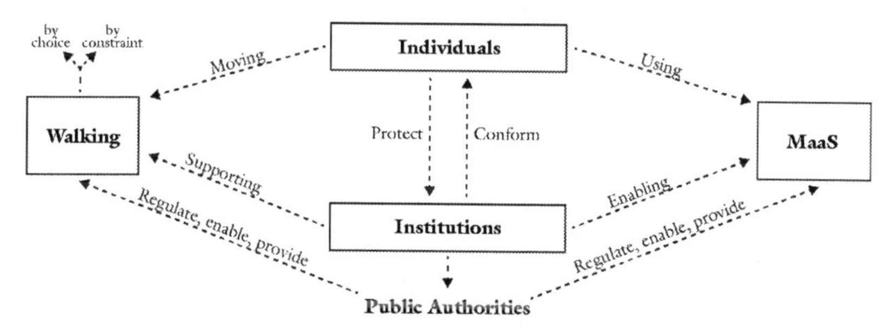

Fig. 4.6 Dynamics of ecosystem actors (*Source* Reyes Madrigal, 2024)

Table 4.3 Examples of social and economic implications amongst the analysed dimension with STEEPL framework

Social	*Economic*
Marginalisation of disadvantaged population	Health impacts with economic consequences
Reinforcement of negative connotations for walking	Unequal access to opportunities for individual and social economic development
Reinforcement of the lack of visibility for walking	Missing opportunites for sustainable value creation
Disregard of diverse lifestyles and needs	Missing opportunites for increasing territorial attractivity
Changes and new health needs	

These aspects are very important since they are linked to the second layer of the digital accessibility notion, the data.

To answer the second question, *how to measure the impacts of the integration of walking in MaaS*, it is required to develop an evaluation framework that takes the complexity of mobility ecosystems into account. Some of the challenges linked to this question are to identify the dimensions to evaluate, the actors that should take that role, and the detailed KPIs to measure. We address these challenges by analysing the impacts of encouraging walking.

4.2.4 *Methodology*

Three methods are used to address the topic of pedestrian mobility integration in MaaS. First, a User Experience (UX) angle is used to look at the accessibility of pedestrian mobility-related information through MaaS' UX/UI using Universal Design Principles (Manley, 2011; Story, 2011; Lynch and Horton, 2016; Brög et al., 2002; Ballard, 2007). Next, we analyse ecosystem dynamics amongst actors to identify the levers and barriers for developing walking in MaaS (Mitchell et al., 1997; Schmeer, 1999; Bunn et al., 2002; Reed et al., 2009). Finally, we use agent-based simulation to evaluate the impacts of providing economic incentives for walking. These incentives are proposed to be passed on to car-drivers, under the logic of 'who pollutes, pays' (Chouaki et al., 2024).

Enhancing User Experience

The overall user experience in MaaS platforms and the data integrated in them depends largely on the choices made by stakeholders shaping the MaaS platforms though their roles such as MaaS operators, providers of technological services, or MaaS developers. These roles can be taken by different actors, however, sometimes there is a single actor in charge of developing and operating MaaS (e.g., when PTAs take the role as MaaS operators and develop an in-house solutions, such as in the Paris region).

To identify the way walking is integrated in MaaS, we identify how information is offered through the MaaS platforms when an itinerary is requested. Our focus is on the information provided for both only walking trips and intermodal walking and public transportation trips to/from and within Paris. The layout of menus and buttons to access the different modes and itineraries were analysed in four MaaS and MaaS-like solutions available in the Paris region. The solutions operated under different governance schemes, including semi-public and public ones led by public transport operators or authorities (e.g., Bonjour RATP or IDFM), as well as the private solutions GoogleMaps and CityMapper. We relied on the universal design principles (Brög et al., 2002; Ballard, 2007; Story, 2011; Manley, 2011), which emphasise creating designs that cater to diverse needs. Additionally, we incorporated basic design principles for the UX, which helped communicate the structure and hierarchy of the information and recommend adapting the placement of elements to effectively convey the priority of the information for the analysis (Reyes Madrigal et al., 2023).

Ecosystem Dynamics Amongst Actors

Aside from the UX, we look at the relationships within the larger ecosystem. A hypothesis is that the little attention paid to walking in MaaS could origin from a lack of awareness of the actors about the benefits of walking if it would be better integrated. A more thorough consideration of walking into MaaS would be a lever for socio-territorial inclusion towards social sustainability, a step against the excessive use of private cars and towards environmental sustainability, a lever for good health, an instrument to know the users and the territories, a field for complex actorial games, and an instrument for regulating uses.

When analysing the stakeholder's dynamics, it is important to delve into the governance dynamics, including the actors involved, the institutional landscape, and the nature of the instruments used (Treib et al., 2007; Hirschhorn et al., 2019). Some of the historical responsibilities of mobility managers encompass working to achieve systemic improvements in performances, efficiency, effectiveness, and equity (Taylor, 2017). Elements like the use of public space, accessibility, density, traffic congestion, environmental impacts, accident prevention, and social inequities have been a big component of the agenda of cities and concerned institutions (Banister, 2008; Allam et al., 2022).

In our research, we address these organisational dynamics through a stakeholder analysis (salience analysis), determining the level of power and interest of stakeholders, through surveys, interviews, and workshops. We aim to identify the power of stakeholders to improve pedestrian mobility in MaaS by looking at levers such as the economic resources, political inferences, decision-making status, potential to develop technological aids for walking, and the interactions with other stakeholders in the ecosystem (Mitchell et al., 1997). Stakeholders are interviewed directly about their level of involvement and interest in the project and their tools in power to influence the outcomes. They are asked to place other stakeholders in one of the four quadrants with high and low variants of the levels of interest and power (Mitchell et al., 1997; Fig. 4.7). The level of power is the ability of a stakeholder to influence the project or to make decisions that affect the project. It can be assessed based on the role, responsibility, resources, and relationships that the stakeholder has with the rest of the ecosystem or the MaaS in question. The level of interest is the degree of involvement or concern of a stakeholder in relation to the project or its outcomes. The level of interest can be evaluated based on the expectations, needs, benefits, risks, or opportunities that the project presents for the stakeholder. To name an example, end users are a key MaaS stakeholder in the ecosystem of actors to whom MaaS is of primary *interest* since they require the services and information offered through/by MaaS solutions. On the other hand, end users may have low *power* because they do not have a significant weight or influence on the project decisions.

Agent-Based Simulations

The third axis of the methodological framework focuses on exploring diverse policies to encourage the adoption of more sustainable mobility practices, with a particular emphasis on behaviour change. Moreover, we aim to understand how such strategies can effectively encourage individuals to choose walking as a transportation mode, how this modal shift would impact the use of public transport and private cars, and ultimately promote sustainable urban mobility. Simulating these incentive policies

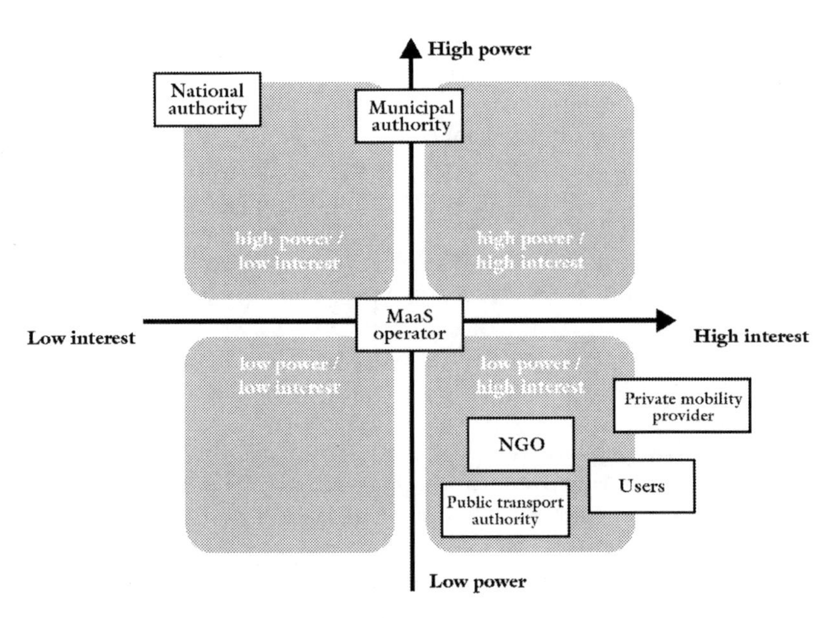

Fig. 4.7 Example of the stakeholder analysis conducted to identify levers for walking in MaaS (*Source* Reyes Madrigal [2023] adapted from Mitchell et al. [1997])

supports the identification of impacts on overall transportation choices, CO_2e emissions, and distances travelled.

In the Île-de-France region, sustainable and active urban mobility options like walking have great potential to promote environmentally friendly transportation and improve public health. However, despite the well-known benefits of walking, it is essential to provide empirical evidence to support its development as a preferred mode of transportation in the eyes of various governmental authorities. To do so, we apply a mixed methods approach, combining a qualitative theoretical and analytical framework and agent-based modelling. The used data is collected from scientific literature, grey literature, mobility surveys, and transport budget databases. The modelling is done via 36 variants, simulated for 100 iterations respectively, resulting from two different financing schemes: polluters pay or external subsidies, as well as six different incentive amounts for the three different types of trips. The simulation results are analysed for modal shift, costs for incentives, CO_2e emissions and their economic value, and specific policy objectives. The results help to answer the questions which policy could bring the best cost-effectiveness, which recommendations can be made, and what future research perspectives arise.

We employ a quantitative modelling approach using agent-based simulations to explore the 36 different incentive variants. In these simulations, we consider individual behaviour and employ a mode choice model that takes into account trip duration, cost, and other relevant factors. By extending a representative simulation of the Île-de-France area and integrating the incentives into the mode choice model, we assess the potential impact of each variant on user decisions (Fig. 4.8).

The research allows us to gain a deeper understanding of the significance of formulating robust and well-defined policy strategies to promote the adoption of alternative transportation modes, regardless of the specific incentive amount offered. The findings of this study offer valuable insights to decision-makers and stakeholders in the region, providing them with the means to implement effective incentive policies across various variants and work towards achieving sustainable objectives.

Most of the proposed incentive strategies have demonstrated their effectiveness in facilitating a shift away from private car usage, particularly evident in the polluter pays scheme (OECD, 1995). These strategies also align with and support various policy objectives on a global, European, and national scale, aimed at combating the impacts of climate change

Fig. 4.8 Simulated incentive policies (*Source* Chouaki et al. [2024]. Image credits: Canva pro [goodstudio/sparklestroke])

and enhancing the overall quality of life on earth. By encouraging new modal shares through well-designed incentive policies, we can contribute to a more sustainable and environmentally friendly mobility landscape in Paris, the Île-de-France region, and beyond.

In general, the incentive strategies that implement the 'polluter pays' component are more effective for discouraging the use of private cars than their counterparts. Moreover, the incentives for walking towards/ from public transports produce a more significant modal shift than the incentives for walking as a sole mode. As expected, the greater the amount of the incentives per kilometre, the more the effects of different strategies differ. These differences can already be noticed with amounts as low as 0.2 €/km.

Regarding the use of public transport, the incentive strategies targeting walks combined with public transport use have an effect on the modal share of public transport. When incentivising walked trips only without the 'Polluter pays' component, the shares of public transport trips slowly decrease as the incentive amount increases. However, with 'Polluter pays', public transport gains users even though the incentive does not directly encourage this mode. This can be explained by the extra cost of using cars that encourages travellers to switch to public transports. This can be seen when comparing the shares of public transport trips according to trip distance across the different variants. The extra cost on the use of car majorly allows to shift long trips to public transport.

4.2.5 *Outcome and Recommendations*

Combining the insights from the three approaches to address the topic of pedestrian mobility in MaaS, we identified several key implications resulting from the current level of integration of walking in MaaS. From the STEEPL categories,[3] the social and economic dimensions were the ones with most potential impact (Table 4.3).

We found that the social and economic challenges identified in the first stage of our research also belonged to the physical dimension of walking, more related to the actual mode share than to its integration to MaaS platforms (see Reyes Madrigal et al., 2023, for a detailed analysis). Knowing this, the agent-based simulation permits us to identify different ways to address walking as an isolated research object. When identifying the impacts of the incentives, we looked at changes in the following indicators: CO_2e emissions, walked distance, and modal share of walking, cycling, public transport, and private car. We chose to analyse these aspects since they are 'variables associated with physical activity interventions' as found by Sallis and Owene (1998). The results show that incentivising walking to reach public transport, both in the polluter pays and subsidies financing schemes, significantly increases the share of public transport trips, as well as the walking trips. This supports raising awareness of the expenditures in pollution-related externalities in transportation. It further sheds light on the fact that other options are possible, such as tackling individual behaviour and promote more sustainable mobility habits. The modal share shifts, especially those resulting from the polluter pays scheme, allowed to identify the framework's capabilities to contribute to the EU's 'fit for 55' goals of decarbonising transportation by increasing the share of active modes and public transport (EC, 2021).

In conclusion, most proposed incentive strategies are effective means to attain a modal shift away from the private car, especially in the polluter pays scheme. They further support various policy objectives across scales, including combatting the climate crisis and maintaining or improving the quality of life. Zooming out again, improved integration of walking in MaaS solutions is an inclusive measure to give individuals equitable access to opportunities in urban, peri-urban, and rural areas and fight against the segregation of territories. Further, it tackles marginalisation as walking

[3] STEEPL framework to analyse Social, Technological, Economic, Environmental, Political, and Legal impacts (Yüksel, 2012).

is the foundation of most mobility chains and the most 'democratic' mode. Better integration of walking is ensuring that MaaS solutions have a positive role in social, environmental, and economic sustainability.

4.3 Comparing Impacts of Mobility Interventions in Metropolitan Cairo

In the last case study, the geographical focus moves south of the Mediterranean Sea to the metropolitan area of Cairo, home to over 22 M people. As in the previous cases, a transdisciplinary approach is used. This time, strategic foresight, personas, and agent-based simulation methods are used to compare the impacts of several urban design and mobility interventions. The urban mobility system of Greater Cairo is simulated for today and for four 2030 scenarios. For the latter, we analyse system impacts of three interventions: Bus Rapid Transit (BRT), better walkability around stations, and improved intermodality. The work results from a collaboration with researchers of the American University in Cairo and Transport for Cairo. The following sections take up the structure of the preceding sections and start by defining the problem, presenting involved stakeholders, the applied methodology, and the anticipated impacts and ways of assessing them. Finally, the outcomes and resulting recommendations are presented.

4.3.1 Problem Definition

The larger challenges of urban mobility apply to most cities, including Cairo. These are the various resource uses, pollution, and challenges for accessing opportunities, unequally distributed across the population. In comparison to Paris, the challenges in Cairo are aggravated by a higher overall population, ongoing urbanisation and population growth, significant levels of poverty, and lack of sufficient resources. These characteristics are common to many large urban areas in developing and least developed countries that will house most of tomorrow's population and complexify any attempts to address the challenges of urban mobility adequately. Specific contextual challenges for Cairo include the hot climate, long distances and often divided functions, limited financial resources to afford specific modes of mobility, significant social and economic costs resulting from traffic congestion and long commutes, as well as increasing environmental pollution, amongst others.

On a methodological side, the focus is on the ability to compare the potential impacts of a set of heterogeneous interventions, including coming from urban design and the transport sector, taking into consideration the complex interactions between them. Further, the focus is again on the heterogeneous user profiles, trying to estimate how different people are affected by specific interventions. Finally, the consideration of future developments is even higher in a context where fast and uncertain developments dominate.

The defined problem is thus the challenge of comparing the cross-dimensional impacts of different solutions considering future uncertainty and disaggregating by user groups. Combined with the costs of interventions—outside of the scope of this section—the resulting information can inform policy and investment recommendations.

4.3.2 Stakeholders

The urban mobility system stakeholders concerned by this section are highly diverse. Similar to Paris, the 22 M people are the key stakeholders together with actors such as public transport authorities and operators, private operators, and regulatory institutions. Further, Cairo being part of the capital region and most populous area of Egypt, the national government has a high interest for the development and prosperity of Cairo. Finally, several international organisations such as the French Development Agency and private companies are involved in current projects and operations. This case study aims to provide a methodological framework to inform policymaking and investment choices to move towards a sustainable and people-centred mobility transition. Thus, it targets primarily national and local government institutions as well as international organisations. However, due to the range of interventions, most actors are concerned in one way or the other.

4.3.3 Methodology

The Cairo case combines different methods. These are qualitative data collection and co-creative methods as well as quantitative modelling and agent-based simulation approaches. The overall methodology can be separated into four sections. First, future scenarios are localised via primarily qualitative methods. Next, a baseline simulation of Cairo is created, and the simulations are adapted to integrate a set of future synthetic populations. Third, three promising interventions are compiled and adapted to the Egyptian context. Finally, the resulting scenarios and their variations are simulated. Each of these steps is described in the following paragraphs.

Scenario Localisation

Building on data from desk research, literature, expert interviews, and site visits, an expert workshop was organised to compile and discuss future trends and uncertainties and their manifestation in Cairo until 2030. A follow-up survey with experts was conducted. This resulted in lists of uncertainties and trends, prioritised by their perceived importance and impact. Using the four future scenario archetypes *Growth, Collapse, Discipline, and Transform* (Dator, 2019), we create narrative descriptions and visualisations of four future scenarios for Cairo 2030. One of them is shown in Box 4.2. The four scenarios are created using the 2*2 matrix, with level of car ownership and sprawl vs. densification as two critical uncertainties.

Box 4.2 Exemplary Cairo scenario narrative and visualisation: Auto-sprawl [method]

Greater Cairo has undergone significant urban growth, accompanied by increased car ownership. Increasing sprawling residential areas have emerged, and the roads are filled with a greater number of vehicles. The government has invested in expanding road networks and implementing intelligent traffic management systems. Whilst car ownership has risen, efforts to promote sustainable transportation include an expanded metro network and bike-sharing programs. Digitalisation plays a crucial role in managing urban growth, optimising traffic flow, and enhancing urban services. Greater Cairo strives to balance private vehicles with sustainable options, ensuring efficient mobility and improving residents' quality of life.

Synthetic Population Scaling

Identical to the first case study, the synthetic population is again the primary factor to adapt the simulations to the four future scenarios. The first step is the creation of a synthetic population for the present before adapting it for each of the scenarios. The geographical study extent was defined in collaboration with mobility experts from Transport for Cairo (TFC) to encompass major populated areas as well as make use of the administrative boundaries of qism, an urban sub-unit under the governate level. The Labour Force Survey by the Economic Research Forum was used and scaled using its statistical weights. Combined with population density data per geographical unit, this allows to scale the population statistically representative to an overall population of about 22.9 M. For the data on mobility behaviour, two surveys conducted by TFC are used (Origin-Destination survey of about 10 k entries and 2.1 k travel interviews). They contain information on what type of people move by which mode, with what trip purpose, and when from which origin to which destination. The travel interviews provide more detailed information such as the number of trips on the previous day. Detailed trip data of both surveys is randomly assigned to daily trip profiles by matching socio-economic data. This results in a database of trip profiles with multiple activities per day. These trip profiles are scaled up and randomly assigned to the overall population. Next, time profiles are created building on purpose-specific trip durations and activity-based start times, taking the number of activities per person into consideration. Start and end times are randomised using a Monte Carlo-based approach. The schedules are assigned randomly to the population based on the number of activities. The resulting movement distribution over a day has been compared to existing traffic peak data for validation. This results in a population of about 15.9 M moving individuals on a normal weekday with a total of 38.2 M daily trips, or on average a bit over 2.4 trips per person per day.

Next, locations for home locations and other activities are added. This process is divided into three parts: First, home locations were generated and assigned to the individuals based on their location. Next, activity locations are generated. Third, activity locations are assigned to each individual based on their home location and the Euclidean trip distances from the survey data. The resulting synthetic population appears to be the most complete so far and builds on well-established methods, detailed databases, uses data that has not been available until very recently and makes partial use of the existing synthetic population generation pipeline

Fig. 4.9 Map of Cairo with simulated traffic counts for 2022 simulation represented by width of links (*Data sources* Roads/water from OSM, Boundary from TFC, Traffic counts generated by authors via MATSim)

developed by Hörl and Balać (2021a). After feeding the synthetic population, the street network, and the transport supply via the GTFS files in existing MATSim pipelines,[4] the current situation can be simulated. To reduce running times of simulations, smaller sample sizes are used (Llorca and Moeckel, 2019; Kagho et al., 2022). Figure 4.9 shows the car traffic counts with a scaled segment width based on the number of vehicles passing in one day. For public transport, about 9% trip segments are made by subway and 91% by bus. The most used line is metro line 1 (4%), followed by metro line 3 (3%) and 2 (2%). These outputs have gone through a local expert-based validation.

After the simulation of the present urban mobility system, the four scenarios are simulated. For each of the scenarios, planned future rail-based mass transit is integrated, including metro extensions, new lines, a

[4] Accessible here: https://github.com/eqasim-org/ile-de-france/tree/cairo (integrated by Sebastian Hörl).

monorail connecting the western and eastern new tows, and a high-speed electric train connecting the same two areas via the lower half of a circle passing central Cairo in the south. The synthetic population is adapted using two uncertainties. One is the distribution of where the population will live by 2030. We distinguish between sprawl and densification, translated into modified average density for each of the 74 qism. Regardless of their spatial distribution, the population is assumed to grow from 22.9 M people today to 25.5 M in 2030. For the densification scenarios, the difference between the maximum density in 2022 and the 2022 density for the respective qism is multiplied by a calculated factor to reach an overall population of 25.5 M. This results in a densification primarily in areas with currently lower densities. For sprawl scenarios, the maximum density is replaced by 20,000 people as potential average target value for less dense developments.

The second parameter to adapt is the rate of car ownership. We replicate the clustering approach developed in Paris with the objective to create four variations of synthetic populations (Gall et al., 2023). The grouping into ten clusters as good balance between accuracy and number of groups permitted to cluster the total present population. From this population, 20% have access to a car in the household (cf. Samaha and Mostofi, 2020). For two of the scenarios, we assume a stable car ownership rate. However, the population increases from 22.9 to 25.5 M, resulting in a significantly higher number of cars. For the two remaining scenarios with an anticipated motorisation percentage increase, we assume a continuing growth of cars per household. We use an average annual growth rate of about 6.9%. Normalising by an annual population growth of +2%, this results in an average increase of 4.9% more cars per capita per year. Assuming this to apply homogeneously to the current rate of 20% car access in the synthetic population, a car access rate of 29.3% is assumed for the two car-centric scenarios in 2030.

Based on a homogeneous population growth for each scenario, the two different spatial population distributions sprawl and densification, and the two car ownership evolvements, we can adapt the original synthetic population. First, the 15.3 M moving people in the current synthetic population are assigned to one of the qism areas based on their home location. Next, for the two scenarios with stable car ownership, the populations are scaled linearly for each qism according to the established changes of the populations. For the remaining two scenarios, the population is adapted to match the target car access rates by using a calculated scaling rate for the people with and without vehicles. This results in one synthetic population for each of the four scenarios, each with about 17.2 M people who take at least one return trip on an average day.

Compilation of Interventions

After the baseline simulation and the adaptation of the synthetic populations for each of the localised future scenarios, interventions are selected to be tested across the scenarios. These origin from the conducted interviews and workshop, as well as existing research and ongoing public discussions. Three interventions are chosen based on their relevance, heterogeneous characteristics, and possibility to both obtain data and model them in the simulation framework.

The first potential solution is a BRT system. Since first developed in Curitiba, Brazil, in 1974, it has been replicated in various cities globally. The main advantages are that it can carry a higher capacity due to optimised boarding and dedicated lanes, often comes with larger busses and higher average speeds, whilst being much cheaper and more flexible than rail-based mass transit options. Figure 4.10 shows a conceptual diagram highlighting the core elements, here for the multi-lane ring road of Cairo. First, there is a dual-direction dedicated lane, usually physically separated to ensure that busses can pass traffic and are not blocked. Second, stations resemble rather tram stops than tradition bus stops. They are often in the middle of the road, accessible via a dedicated pedestrian infrastructure and with a ticket purchase and fare collection system at the station instead of inside the bus. Usually, the platforms are raised, and multiple doors permit rapid disembarking and boarding (see also pre-feasibility study by ITDP, 2015). A potential route of a bit under 100 km with 46 stops on the ring road surrounding most of central Cairo and passing New Cairo in the east has been defined by TFC.

The second solution origins from the discipline of urban planning and design. Mass transit success is impacted largely by the catchment area of stations as it defines how many people can reach a station. In practice, this is defined at three levels (Fig. 4.11). First, the urban form defines the reachability. As smaller the blocks are, as more walkable they are (cf. Jacobs, 1961). Second, the quality of the walking infrastructure impacts

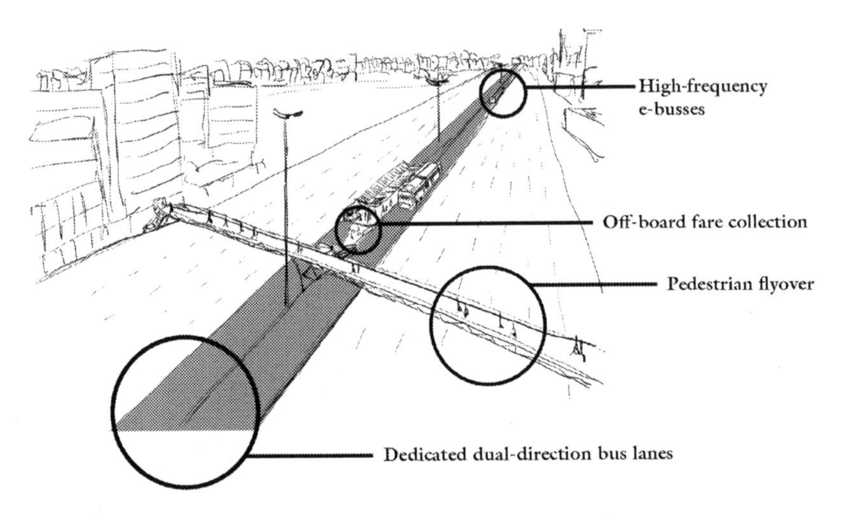

Fig. 4.10 Bus rapid transit on Cairo ring road concept diagram

the likelihood for some or all user groups. Third, a variety of other parameters, such as shading, perceived security, passing interesting/high-quality spaces, or the co-location of other waypoints like grocery shops or a postal office can impact the likelihood of people to walk.

The objective is not to specify which type of intervention could be implemented across all Greater Cairo. Instead, the objective is to measure how a set of place-specific measures that increase the walkability everywhere could have an impact on the overall mobility system. For this, we turn to the choice model of MATSim. Every agent attempts to maximise her individual utility (Horni et al., 2016). The utility is a function of primarily time and cost within the constraints of available options (e.g., car availability) (Hörl, 2023; Balać et al., 2023). In more detail, different values are integrated, such as travel time, parking pressure, headway, in-vehicle time, waiting time, per km or standardised public transport costs (Hörl, 2023). With this basis, a utility is calculated for each agent. It is not possible to have a positive score resulting from walking. Instead, there is a linear relationship between the distance of the station and the score. The farther the distance, the lower the score. In practice, this means that depending on the availability of other options, at a certain distance, the negative impact on the utility of walking to the station outweighs the advantage of the public transport option. This cut-off point is in

Fig. 4.11 Three aspects impacting walkability (left: urban form, centre: infrastructure availability and quality, right: quality of public space)

planning practice often set at 400–450 minutes, the distance it takes on average 5 minutes to walk. In reality, this value is often higher depending on transport supply, alternatives, climate, or social practices. For the second intervention, we assume an improvement of 15% compared to the current situation. A stated preference survey could permit to quantify the monetary value of an improvement further.

The third and final intervention relates to smart and sustainable cities, MaaS, and transport data management. Inter- and multimodal transport is often mentioned as important component of sustainable urban mobility systems (Oostendorp and Gebhardt, 2018). Intermodal transport refers to 'transportation by more than one form of carrier during a single journey' (Goetz, 2009) whilst multimodality is defined as the 'use of at least two modes of transportation – bicycle, car, or public transportation – in 1 week' (Nobis, 2010). Public transport users are more likely to use one bus or metro even if it takes slightly longer than a combination of multiple modes (cf. Oostendorp and Gebhardt, 2018). This can result from the disturbance of on-board activities, risks of delays or prolonged waiting times, or comfort reasons. For this reason, intermodal exchanges (e.g., bus to metro) are currently integrated in the simulation's utility function as a penalty term. This has particular relevance in Cairo where 'focus group findings also indicate that passengers actively avoid transfers [...] which is likely to reflect uncomfortable, disorganised and unsafe interchange environments as well as service unreliability which together cause transfer stress' (TFC, 2021).

Different types of information can improve the willingness to transfer between modes and use intermodal trips and thus strengthen an important component of sustainable urban mobility transitions. One example is the improvement of the physical transfer point. This can include more developed stations and lighting, shading, camera surveillance, or other protective measures. The transformation and adapted management of the space around stations can lead to improved intermodality, e.g., by having a dedicated area outside of a metro station where buses can be accessed directly and safely. Third, the access to up-to-date information is crucial. If somebody knows that the connecting mode will arrive in a few minutes and can rely on it, she will be more likely to consider it as an option. This requires a combination of transport data management, sharing between providers, digitalisation, in general, across modes, as well as different information access points, such as via smartphone apps, existing mobility providers, and information panels. Again, we do not intend to suggest a single solution for every situation but instead aim to quantify the potential impact an improvement of intermodality could have at the system level. For this, like before, we assume an improvement by 15% of the value used in the baseline simulation, meaning a reduction of the penalty score for intermodal transfers (cf. Nagel et al., 2016).

Agent-Based Simulation

In the last step, the four developed scenarios for 2030 with their respective synthetic populations and the three considered solutions can be modelled. A key potential of agent-based simulations compared to other models is the consideration of systemic impacts of the choices of other agents and the interplay of different services. Hence, we not only simulate the scenarios with each intervention separately but also with all different combinations. This results in one base simulation of the present, four simulations of the contextual scenarios for 2030 without interventions, twelve simulations for each intervention individual across the scenarios, twelve simulations combining each time two solutions across scenarios, and four simulations where the combination of all three intervention is tested together. This results in a set of 33 simulations. Each of them is simulated with identical settings except replacing the synthetic populations as described in the previous section and adding the BRT transport supply as well as the changed values in the utility function for the other two interventions.

4.3.4 *Impacts and Assessment*

Lastly, we zoom out again and remember the underlying problem. The objective is providing a method that permits assessing and comparing potential solutions whilst integrating future uncertainty and paying particular attention to impacts on different groups of people. To focus on the comparative character, we provide normalised per capita values for five representative indicators, each time for the last simulated iteration. The score refers to the calculated utility. CO_2e is calculated by the number of car vehicles multiplied 150 g CO_2e/pkm. The time is the total amount of time spent per day and person. The walking indicator results from the percentage of walked trips for those shorter than 1 km. The public transport indicator shows the modal share for trips longer than 1 km. CO_2e and time are both reversed to have homogeneously more positive values as higher the value is. Figure 4.12 (left) shows the comparison of the 2022 situation (transparent grey) with the four scenarios. Scenario 1 'Autosprawl' has clearly the lowest values, followed by Scenario 2 'Gridlock'. The Scenario 2 'Disconnection' has the highest score in walking, most likely resulting from the fact that low density and few cars result in the necessity for most people to walk regardless of trip length. Scenario 4 'Urban revival' performs similar to 2022 as the only 2030 scenario and clearly outperforms the other future scenarios. On the right side of the same figure, for each of the future scenarios, the base option (future transport supply and population) and the option with BRT, walking, intermodality are compared. Two findings are noteworthy. First, the worse the scores are, the higher the impact of the interventions. The two scenarios with a car ownership rate increase are benefitting most. On the other hand, the Gridlock scenario achieves an identical public transport rate for trips over 1 km, highlighting the potential to shift people to public transport even if more cars are there as long as the population density permits the access to public transport.

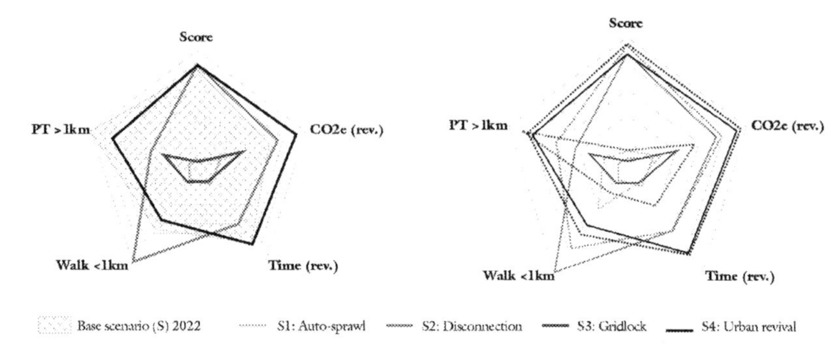

Fig. 4.12 Left: Comparison of 2022 scenario and four 2030 via five indicators. Right: Comparison between base scenario (continuous line) and option d with BRT + walking + intermodality for each of the four 2030 scenarios

If we take each of the five indicators with an equal weight, the 'Auto-sprawl' scenario performs the worst and the 'Urban revival' the best. To analyse the impact of the different interventions further, we zoom in on these two scenarios. The base option is relatively homogeneously underperforming across the five indicators. The other scenario options are performing slightly better across all categories except the time spent. However, only the option combining BRT, improved walking to stations, and intermodality performs significantly better in general. On the other hand, the impacts are less clear for the scenario 'Urban revival' where the option without changes does not perform worse than the base scenario. This could be interpreted by a lower sensitivity to the walkability and intermodality indicators as a more densified urban form permits to walk to stations in any case or alternatively reach more places close by. The BRT system on the ring road is less relevant as more people live centrally and have access to other modes of transport. Other analyses, such as those conducted for the first case study can be conducted but are outside of the scope of this section.

4.3.5 Outcome and Recommendations

This case was chosen to showcase the potentials of a mixed methods methodological framework. This refers to scenario-based approaches, their co-creative localisation and transformation through trends, uncertainties, and archetypes, as well as the application of the urban mobility system model and rendering them tangible via data-driven personas and synthetic populations. The approach enabled in-depth insights in the local system dynamics and has received positive reactions throughout and permitted to jointly elaborate whilst taking different perspectives on the complex challenge of transforming urban mobility systems.

The results are aiming at a higher-level perspective but permit pointing in some directions that can inform some recommendations. First, and on a larger level, Cairo—at least as much as most other metropolitan areas—is highly complex and fully of uncertainties. The consideration thereof in policymaking and project planning appears crucial. Similarly, the use of already available data permits more detailed and systemic understandings of the intricacies of Cairo. Regarding the scenarios, simulation results highlight the significant and negative impact of the scenarios with increasing car ownership rates. Current development patterns of continuously adding lanes and road infrastructure will not resolve the already existing gridlock-like situation. Instead, recent investments in public transport of various types point in a direction of more effective, sustainable, and equal mobility supply of tomorrow. The three tested interventions all show potential positive impacts, individually and even more so if applied jointly. Whilst some interventions require significant infrastructure investments, some others, such as improved data availability and station management, might have significant impacts as well for much less costs. Lastly, whilst all tested interventions, as well as the planned future transport supply seem to improve the situation, the strongest impact origins from the spatial layout and car ownership. Whilst the latter can be understood as a complex consequence of needs, resources, and social practices, the former is in the hands of the public sector and the planning bodies. The ongoing sprawl, low density and gated community trends, resulting in increasingly disconnected urban fabric, challenges (amongst many others) any potential pathway to sustainable urban mobility futures. A policy recommendation is therefore—within limitations of the applied method—the setting of an enabling foundation of the spatial structure, followed up on by a mix of mass transit infrastructure and supporting interventions.

A few limitations and future work potentials must be mentioned. First, scenarios should always be adapted to the place-specific needs and regularly updated. We aimed at providing a method and first set of scenarios but highlight the need to refine and extend them continuously. On the simulation side, several smaller limitations and potential for future work have been already mentioned throughout this sub-chapter. The most important ones are the further detailing of the synthetic population as well as ongoing calibration to match the observed situation whenever more or higher-detail data becomes available. Further, a set of local specificities that are highly relevant for Cairo's mobility system are both promising as way to improve the model accuracy as well as, in some cases, interesting research topics in themselves. This includes dominant features such as wide-spread one-way streets in many central areas since a couple of years which could be—with the right dataset—easily be integrated and are assumed to improve the simulation quality mostly in central Cairo. Further, paratransit and semi-formal on-demand mobility modes that exist in Cairo as well as many other fast urbanising cities could be integrated. This would allow a higher simulation quality, permit to estimate, for example, the impact of the electrification of two- and three-wheelers as currently starting and ongoing across cities in Africa and Asia in relation to its contribution to sustainable and people-centred mobility, as well as providing replicable components for the MATSim framework that could lead to a growing methodological uptake in countries with more heterogeneous mobility modes. Further, more detailed surveys or other data, such as mobile phone data, that permit the detailing of the trip chain schedules would be highly beneficial. Along the same lines, an improved understanding of the local value of time through targeted surveys would permit calibrating the underlying choice model. In Cairo, even more than in Paris, a particular potential lies on understanding and integrating the differences between socio-economic groups as currently even many public transport options remain unaffordable for a significant part of the population (TFC, 2021). Lastly, as soon as more accurate and better calibrated simulations are possible, it would be worthwhile to run more sensitivity tests across scenarios and interventions, as well as to scale up the sample sizes of the simulations for more local analyses or the integration of dynamic modes such as tuk-tuks or shared on-demand taxis without fixed routes.

References

Allam, Z., Bibri, S., Chabaud, D., and Moreno, C. (2022) The theoretical, practical, and technological foundations of the 15-minute city model: Proximity and its environmental, social and economic benefits for sustainability. *Energies*, Vol. 15. https://doi.org/10.3390/en15166042.

Babst, V., Du Bernard, C., Gangloff, C., and Kautzmann, J.-E. (2022) Results based management. Approach of the Council of Europe. Practical Guide. Council of Europe. Directorate of Programme and Budget Programme division.

Ballard, B. (2007) *Designing the mobile user experience*, 1st ed. Wiley. https://doi.org/10.1002/9780470060575.

Balać, M., Hörl, S., and Basil, S. (2023) Discrete choice modeling with anonymized data. *Transportation*. https://doi.org/10.1007/s11116-022-10337-1

Banister, D. (2008) The sustainable mobility paradigm. *Transport Policy*, Vol. 15/2, pp. 73–80. https://doi.org/10.1016/j.tranpol.2007.10.005.

Bertraud, A. (2018) *Order without design. How markets shape cities*. Cambridge/London: The MIT Press.

Brög, W., Erl, E., and Mense, N. (2002) Individualised marketing changing travel behaviour for a better environment. http://www.epomm.eu/newsletter/v2/content/2015/1115/doc/IndiMark.pdf.

Bunn, M.D., Savage, G.T., and Holloway, B.B. (2002) Stakeholder analysis for multi-sector innovations. *Journal of Business & Industrial Marketing*, Vol. 17(2/3), pp. 181–203. https://doi.org/10.1108/08858620210419808.

Chatziioannou, I., Nikitas, A., Tzouras, P.G., Bakogiannis, E., Alvarez-Icaza, L., Chias-Becerril, L., Karolemeas, C., Tsigdinos, S., Wallgren, P., and Rexfelt, O. (2023) Ranking sustainable urban mobility indicators and their matching transport policies to support liveable city Futures: A MICMAC approach. *Transportation Research Interdisciplinary Perspectives*, Vol. 18/100788. https://doi.org/10.1016/j.trip.2023.100788

Chouaki, T., Reyes Madrigal, L.M., and Hörl, S. (2024) Assessing the impact of monetary incentives for walking using agent-based mobility simulations and discrete mode choice models. TRB Annual Meeting. Washington DC, USA, 07–11 January 2024 [forthcoming].

Dator, J. (2019) What futures studies is, and is not. In: *Jim Dator: A noticer in time. Anticipation science*, Vol. 5, Cham: Springer. https://doi.org/10.1007/978-3-030-17387-6_1.

Dong, L., Santi, P., Liu, Y., Zheng, S., and Ratti, C. (2022) The universality in urban commuting across and within cities. https://doi.org/10.48550/arXiv.2204.12865.

European Commission (EC). (2021) Delivering the European Green Deal. https://commission.europa.eu/strategy-and-policy/priorities-2019-2024/european-green-deal/delivering-european-green-deal [accessed August 2023].

Gall, T., Chouaki, T., Vallet, F. and Yannou, B. (2023) Un cadre basé sur les scénarios du futur au service de la simulation multi-agents de la mobilité urbaine de demain. s.mart colloque, Carry-le-Rouet, 04/2023.

Geurs, K. and van Wee, B. (2004) Land-use/transport Interaction Models as Tools for Sustainability Impact Assessment of Transport Investments: Review and Research Perspectives. *European Journal of Transport and Infrastructure Research*, Vol. 4/3, pp. 333–355.

Goetz, A.R. (2009) Intermodality. In: International Encyclopaedia of Human Geography.

Hirschhorn, F., Paulsson, A., Sørensen, C.H., and Veeneman, W. (2019) Public transport regimes and mobility as a service: Governance approaches in Amsterdam, Birmingham, and Helsinki. *Transportation Research Part A: Policy and Practice*, Vol. 130, pp. 178–191. https://doi.org/10.1016/j.tra.2019.09.016.

Hörl, S., and Balac, M. (2021a) Synthetic population and travel demand for Paris and Île-de-France based on open and publicly available data. *Transportation Research Part C: Emerging Technologies*, Vol. 130, p. 103291. https://doi.org/10.1016/j.trc.2021.103291.

Hörl, S., and Balac, M. (2021b) Open synthetic travel demand for Paris and Île-de-France: Inputs and output data. *Data Brief*, Vol. 39, p. 107622. https://doi.org/10.1016/j.dib.2021.107622.

Hörl, S. (2023) Towards replicable mode choice models for transport simulations in France. *9th International Symposium on Transportation Data & Modelling (ISTDM2023)*, 19–22 June 2023, Ispra.

Horni, A., Nagel, K., and Axhausen, K.W. (eds.). (2016) *The multi-agent transport simulation MATSim*. London: Ubiquity Press. https://doi.org/10.5334/baw.

Horschutz Nemoto, E., Issaoui, R., Korbee, D., Jaroudi, I., and Fournier, G. (2021) How to measure the impacts of shared automated electric vehicles on urban mobility. *Transportation Research Part D: Transport and Environment*, Vol. 93, p. 102766. https://doi.org/10.1016/j.trd.2021.102766.

ITDP (2015) Bus rapid transit for Greater Cairo: Prefeasibility assessment. *Institute for Transportation and Development Policy*, June 2015.

Jacobs, J. (1961) *The death and life of Great American cities*. New York (NY): Random House.

Kagho, G.O., Meli, J., Walser, D., and Balac, M. (2022) Effects of population sampling on agent-based transport simulation of on-demand services. *Procedia Computer Science*, Vol. 201, pp. 305–312. https://doi.org/10.1016/j.procs.2022.03.041

Knapskog, M., Hagen, O.H., Tennøy, A., and Rynning, M.K. (2019) Exploring ways of measuring walkability. *Transportation Research Procedia*, Vol. 41, pp. 264–282. https://doi.org/10.1016/j.trpro.2019.09.047.

Llorca, C. and Moeckel, R. (2019) Effects of scaling down the population for agent-based traffic simulations. *Procedia Computer Science*, Vol. 151. pp. 782–787. https://doi.org/10.1016/j.procs.2019.04.106/

L'usine nouvelle. (2018) Mobike devient le quatrième opérateur de vélos en libre-service dans Paris. https://www.usinenouvelle.com/article/mobike-devient-le-quatrieme-operateur-de-velos-en-libre-service-dans-paris.N642938 [accessed August 2023].

Lynch, P., and Horton, S. (2016). *Web style guide: Foundations of user experience design*, 4th ed. Yale University Press.

Manley, S. (2011). Chapter 17. Creating an accessible public realm. In: Smith, K.H., and Preiser, W.F.E. (eds.), *Universal design handbook*, 2nd ed. McGraw-Hill.

Marchetti, C. (1994) Anthropological invariants in travel behavior. *Technological Forecasting and Social Change*, Vol. 47/1.

Mitchell, R.K., Agle, B.R., and Wood, D.J. (1997) Toward a theory of stakeholder identification and salience: Defining the principle of who and what really counts. *The Academy of Management Review*, Vol. 22/4, pp. 853–886. https://doi.org/10.2307/259247.

Nagel, K., Kickhöfer, B., Horni, A., and Charypar, D. (2016) A Closer Look at Scoring. In: Horni, A, Nagel, K and Axhausen, K.W. (eds.) *The Multi-Agent Transport Simulation MATSim*, pp. 23–34. London: Ubiquity Press. https://doi.org/10.5334/baw.3

Nobis, C. (2010) Multimodality: Facets and Causes of Sustainable Mobility Behavior. Transportation Research Record: Journal of the Transportation Research Board. https://doi.org/10.3141/2010-05

OECD. (1995) Environmental principles and concepts (OCDE/GD(95)124). Organisation for Economic Co-operation and Development (OECD). https://one.oecd.org/document/OCDE/GD(95)124/En/pdf.

Oostendorp, R. and Gebhardt, L. (2018) Combining means of transport as a users' strategy to optimize traveling in an urban context: empirical results on intermodal travel behavior from a survey in Berlin. *Journal of Transport Geography*, Vol. 71, pp. 72–83. https://doi.org/10.1016/j.jtrangeo.2018.07.006

Reed, M. S., Graves, A., Dandy, N., Posthumus, H., Hubacek, K., Morris, J., Prell, C., Quinn, C.H., and Stringer, L.C. (2009) Who's in and why? A typology of stakeholder analysis methods for natural resource management. *Journal of Environmental Management*, Vol. 90/5, pp. 1933–1949. https://doi.org/10.1016/j.jenvman.2009.01.001.

Reyes Madrigal, L.M., Nicolaï, I., and Puchinger, J. (2023) Pedestrian mobility in mobility as a service (MaaS): Sustainable value potential and policy implications in the Paris region case. *European Transport Research Review*, Vol. 15/13. https://doi.org/10.1186/s12544-023-00585-2..

Reyes Madrigal, L.M. (2024) Walking in Mobility as a Service (MaaS): action levers for enhancing the integration of walking in MaaS from a stakeholder's ecosystem approach. Doctoral Dissertation, University Paris-Saclay.

Sallis, J.F., and Owen, N. (1998) *Physical activity and behavioral medicine*. SAGE Publications.

Schmeer, K. (1999) Stakeholder analysis guidelines. https://dev2.cnxus.org/wp-content/uploads/2022/04/Stakeholders_analysis_guidelines.pdf [accessed August 2023].

Story, M.F. (2011) Chapter 4. The principles of universal design. In: Smith, K.H., and Preiser, W.F.E. (eds.), *Universal design handbook*, 2nd ed. McGraw-Hill.

TFC (Transport for Cairo, 2021) Greater Cairo Region Mobility Assessment and Public Transport Improvement Study. Mobility Assessment and Data Collection Report.

Taylor, B.D. (2017) The geography of urban transportation finance. In Giuliano, G., and Hanson, S. (eds.), *The geography of urban transportation*, 4th ed. New York/London: The Guilford Press.

Thomas, L.D.W., and Ritala, P. (2022) Ecosystem legitimacy emergence: A collective action view. *Journal of Management*, Vol. 48/3, pp. 515–541. https://doi.org/10.1177/0149206320986617.

Treib, O., Bähr, H., and Falkner, G. (2007) Modes of governance: Towards a conceptual clarification. *Journal of European Public Policy*, Vol. 14/1, pp. 1–20. https://doi.org/10.1080/13501760601071406.

Vosooghi, R., Puchinger, J., Jankovic, M., and Vouillon, A. (2019) Shared autonomous vehicle simulation and service design. *Transportation Research. Part C, Emerging Technologies*, Vol. 107, pp. 15–33. https://doi.org/10.1016/j.trc.2019.08.006.

Vosooghi, R., Puchinger, J., Bischoff, J., Jankovic, M., and Vouillon, M. (2020) Shared autonomous electric vehicle service performance: Assessing the impact of charging infrastructure and battery capacity. *Transportation Research Part D: Transport and Environment*, Vol. 81. https://doi.org/10.1016/j.trd.2020.102283.

Xenou, E., Ayfantopoulou, G., Royo, B., Tori, S., and Mazzarino, M. (2021) A conceptual assessment framework for capturing the sustainability impacts of city-specific future mobility scenarios: the SPROUT approach. 10th International Congress on Transportation Research. *Future Mobility and Resilient Transport: Transition to innovation*, 2021.

Yüksel, I. (2012) Developing a multi-criteria decision-making model for PESTEL analysis. *International Journal of Business and Management*, Vol. 7/24. https://doi.org/10.5539/ijbm.v7n24p52.

CHAPTER 5

Reflections on Sustainable Urban Mobility Futures

Abstract In the fifth chapter, we discuss the key ideas of the book and point in five directions of future research with perceived high potentials. We pose the questions: What to work towards? How to integrate uncertainty across methods and mental frameworks? How to transfer economic value between actors? How to transform research funding sources? And how to establish accessible modelling standards? Together, these five questions refer to ongoing and planned work of the authors as well as acting as invitation to others to contribute to one question or the other.

Keywords Future research · Normative · Uncertainty · Interdisciplinarity · Value transfer · Alternative funding

This book set out to provide a transdisciplinary overview of urban mobility, its challenges, involved disciplines, and promising already established or arising methods. Using a systems perspective and delineating the concept of urban mobility geographically and conceptually, a couple of leading future developments and trends were introduced. These included sections dedicated to societal, urban, and technological trends, aiming to outline the context and fields of relevance for urban mobility. Next, the focus shifted to approaches that are used to design and plan urban mobility solutions. Starting with a focus on urban governance, the focus

T. Gall et al., *Sustainable Urban Mobility Futures*, Sustainable Urban Futures, https://doi.org/10.1007/978-3-031-45795-1_5

was on high level and enabling supranational, national, and local policy-making and their agendas for and impacts on urban mobility. Next, we looked at people-centred design, what approaches and concepts it entails and how it can be used for designing urban mobility solutions. Lastly, a focus was on data-driven design and decision-making, introducing various concepts and methods on data collection, preparation, modelling, and simulation.

Combing the conceptual and methodological foundations, we explored three case studies which aim all in different ways to contribute to a transition towards more people-centred and sustainable urban mobility. The first looked at the context-specific and people-oriented assessment of novel mobility solutions whilst taking into consideration future uncertainty. The next explored the role of active mobility in new solutions such as MaaS and approaches to estimate and influence the behaviours of people. The last one looked at the comparison of several mobility interventions in Cairo. Each of them combined qualitative and quantitative approaches and methods and originated out of a collaboration between researchers from different backgrounds. The three case studies aimed to showcase how the theoretical concepts can be translated into practical methods whilst providing some more detailed insights in the local contexts and selected potential urban mobility interventions. With this, we aimed to introduce different lenses on urban mobility as well as highlighting its relevance and social, environmental, and economic challenges and potentials. Whilst having attempted to zoom in and provide some in-depth information on some of the topics, we are aware of the limitations of the scope of this book. In most cases, references to further work are provided to permit delving deeper in one topic or the other—whilst keeping the others and their intertwined and systemic character in mind. On the other side, many questions remain open and potentials for future research can be found in each topic as well as on the transversal scale. In the remainder of this chapter, we present five topics that we perceive as relevant to be investigated further and miss sufficient research up to now. These can be seen as topics to reflect on further or as invitation for researchers, students, and practitioners to go further and look for answers within and—more importantly—across domains.

5.1 What to work towards?

The first is at a higher level and related to the notion of *wicked problems*, referring to challenges that have multiple dimensions and no clear solution or goal setting, mostly also within stakeholder ecosystems with diverging interests. At a higher level, common objectives are easy to establish. For example, few would challenge that urban mobility should optimally not pollute, not lead to accidents, and allow people to effectively get from A to B. Two questions are, however, how to balance them as they are today impossible to be achieved in most contexts, and what approaches to use. For example, large polycentric metropolitan areas with sub-centres and fast trains or medium-sized cities where everything can be easily reached? Large investments in public transport and active mobility or shared automated and electric vehicles? It is unlikely that many clear answers to this will be found. However, significant potential lies in identifying, analysing, and accompanying best practices across socio-cultural contexts to understand how to enable and finance sustainable urban mobility transitions, as well as advancing with prospective multi-dimensional assessment frameworks that permit a better understanding of the status quo, possible developments, and the potential impact different solutions might have. This could be done in combination with smart cities, digital twins, or other kind of dashboards, as well as governance and policy standards at national or supranational level (cf. Bertraud 2019). Possible research questions are: How to compare and choose between options of different fields? How to standardise the modelling of urban mobility solutions' impact? How to value social and environmental gains in urban mobility systems?

5.2 How to integrate uncertainty across methods and mental frameworks?

The second question origins from the larger context of possible futures in planning and design. So far, uncertainty finds restricted application in practice despite the well-established existence of methods to do so. Barriers include the resource intensity of foresight, the challenge to integrate unknown, subjective information into quantitative and positivist approaches, and the mental challenge that working with uncertain futures comes with. Nevertheless, at a time where changes are rapid and challenges urgent, any attempt to make policies, designs, and plans of today more 'futureproof' must be exploited. Whilst having made some proposals for this within this work, plenty other methods, frameworks, and fields of application in the context of urban mobility can be explored, spanning qualitative and quantitative approaches, and especially mixed methods approaches. Possible research questions include: Which foresight methods can support urban mobility system design? How to integrate uncertainties of type x in method y? What are the long-term impacts of working with foresight approaches? How to build capacity for uncertainty in urban mobility system design?

5.3 How to Transfer Economic Value Between Actors?

The third question enters the economic dimension which was not at the core of this work but repeatedly referred to. The mobility sector entails enormous costs and benefits. This includes the oil and automotive industry, public transport, the value of time lost in commute and congestion, or the healthcare cost resulting from sedentary lifestyles enabled by car-based societies. In this book, we provided the example of challenges and potentials of better integrating active mobility in MaaS as traditionally free service. However, this can be extended to multiple fields, including pollution, space consumption for parking, or politically laden topics such as generalised costs resulting from peri-urban or rural lives or hypermobile people compared to others. It is unlikely that everyone will pollute and use mobility services equally. On the other hand, few solutions exist so far to redistribute costs. Some toll, tax, parking, or fee systems try to do so but do not reach many fields. More research is needed in understanding the actual value of all key components of the mobility ecosystem and how it might be redistributed. Some possible research questions are: How to estimate the economic value of x? What policies/incentives/penalties have what impact on, e.g., mode shift? What potentials for more active mobility bear insurances/employer schemes? How to match taxation with induced costs at individual level? What potential exists for voluntary schemes or awareness campaigns on economic dimension?

5.4 How to transform research funding sources?

Remaining in the economic dimension, some funding opportunities with systemic foci exist, such as the EU-funded Driving Urban Transitions programme.[1] However, most funding of research in the mobility sector is currently tilted towards technological solutions. This is supported both by the private sector looking for new fields of business development as well as national and supranational bodies that aim to enable technological progress, competitiveness, and attractivity, or in case of EV batteries, even geopolitical independence and *de-risking* value chains by investing in rare material mining, battery, and component construction, as well as technological advancements. In the past years, this led primarily to investment in smart cities, research on EVs and their charging networks, as well as all types of automated vehicles. To a lesser degree but still significant, public and private funding went into urban air mobility, drones, and robots for human mobility and logistics. This results in overrepresented funding of high-tech solutions with—as shown in two of the case studies—limited actual potential und underfunding of active mobility, behavioural measures, or even urban planning as foundation of urban mobility demand. Whilst no precise research questions shall be formulated for this section, new funding sources, increasing public funding, or other models could be explored to, on the one hand, move towards an equal distribution of research foci in the mobility sector, whilst simultaneously strengthening the independence of domain research. Novel funding sources might origin from areas where, for example, active mobility, create value, and therefore an interest exists to understand them better.

5.5 How to establish accessible modelling standards?

The last field of potential future research focuses on the technical dimension, especially mobility modelling. Many modelling approaches and frameworks exist but are so far limited by multiple barriers. These include limited access to data, largely non-standardised data formats, unaffordable modelling software, computational resource intensity, 'black box' approaches, or complex technical approaches that are inaccessible for non-experts. Each of the barriers opens a large field of promising

[1] https://dutpartnership.eu [accessed August 2023].

research. This is extended by a large range of improvements within modelling frameworks such as MATSim as mentioned throughout this book. For example, better approaches for model validation and possible calibration, more detailed choice models with other factors, and heterogeneous choice models to better represent different user profiles. On the other hand, different modelling approaches bear different potentials. For example, integrated approaches combining such methods and permitting the simplified use of parameters from a complex simulation in simpler and faster simulation would allow to scale up and simultaneously scale up more detailed and disaggregated modelling whilst increasing accuracy and democratise more complex modelling approaches. Possible research questions are: How can heterogeneous choice models be calibrated and integrated in modelling approaches? What utility function components bear the potential to improve modelling accuracy? How to validate/calibrate simulations more precisely? What data/modelling/documentation standards can advance open and collaborative mobility modelling?

Reference

Bertraud, A. (2019). *Order Without Design. How Markets Shape Cities*. Cambridge/London: The MIT Press.

CHAPTER 6

Conclusion and Perspectives

Abstract Finally, the last chapter summarises what the larger ambition of the book has been and how we attempted to address it. To emphasise what pathways towards sustainable urban mobility futures require, we reiterate five key concepts and detail the intentions behind them. The five concepts are uncertainty, transdisciplinarity, people-centredness, applied research, and open ecosystems.

Keywords Uncertainty · Transdisciplinary · People-centred · Applied research · Open research

In this book, we focused on sustainable urban mobility futures. The sub-title further promised the discussion of transdisciplinary transition pathways and design approaches. Through a broader framing, a compilation of future trends and developments, multi-disciplinary methods, and three case studies, we attempted to outline what is at stake, what solutions or approaches already exist, and where more work is needed. Without claiming completeness in any of the sub-sections, we hope to have provided an overview on urban mobility from different angles and through various lenses. All presented methods and cases are connected by their ambition to enable transitions towards more people-centred and sustainable urban mobility futures. With emphasis on the specific,

T. Gall et al., *Sustainable Urban Mobility Futures*, Sustainable Urban Futures, https://doi.org/10.1007/978-3-031-45795-1_6

multi-dimensional objectives such as emissions reduction or accessibility increase, as well as the plural of futures due to a multitude of possible pathways between which must be navigated—each with their respective strengths but also weaknesses: A key characteristic of wicked problems. Aside from the research potentials discussed in the previous section, we can summarise the key messages of this book in five points.

Uncertainty has always been around but the complexity and pace of today's world, further aggravated by the climate crisis, require a systematic integration thereof. Whilst no perfect methods exist, many tangible ways are around that enable to consider them partially and simultaneously develop the capacity of users and decision-makers that our assumptions of today are subject to error, but that partial preparation is possible and most certainly better than not doing so.

Transdisciplinary approaches, regardless of in practice or research, are challenging and take more resources to set up in the beginning. However, once a common ground, language, and theoretical framework is established, it permits to find novel solutions to old problems by looking left and right of the disciplinary focus. Problems in fields such as urban mobility cannot be resolved or even effectively addressed from one perspective individually. Thus, aside from its innovation potential, transdisciplinarity is a key ingredient for addressing the pressing challenges of complex urban systems.

People-centredness is the third key message. People are neither homogeneous goods to be transported nor simple mobility service customers to be convinced. Instead, we need to extend the focus towards a heterogeneous concept with diverse needs and requirements. This constitutes the foundation for sustainable mobility *for all* whilst also contributing to mobility justice. The larger focus on people includes further those indirectly affected by, e.g., rare material mining, as well as future generations negatively impacted from the climate crisis. Their consideration in today's conceptualisation is a critical element of people-centred and sustainable solutions for tomorrow.

Applied research and innovation projects are another high-impact lever. This potential has been identified by the European Commission and is implemented through mission-based funding, as well as by increasing collaborations between multiple stakeholders across levels. Whilst the dominant collaboration with the private sector bears risks of tilted research foci, it is a crucial element as most solutions and technologies origin from industry. This is even more so the case when the public sector is simultaneously involved and permits to negotiate between effectiveness, economic and technological viability, as well as fulfilment of social and environmental objectives.

Open ecosystems are the final ingredient. This includes urban governance enabling or enforcing open data practices, the development of data and sharing standards, as well as the advancements on open-source solutions. The combined use of them enables the development of new solutions and methods but also the further upscaling of testing and validating existing methods. Public, private, and academic stakeholders can all contribute to this. The public sector has the chance to create open data policies for actors partaking in projects with public interest such as urban mobility. Further, it can make own data available through open data platforms as increasingly the case, for example, in France. Even for the private sector, the benefits of open data sharing can often outweigh the costs and risks. Transport for Cairo, a contributor to the third case study, is a private consultancy but hosts an extensive and growing open data platform, whilst the research institution hosting the Anthropolis Chair, IRT SystemX, as well as the academic partner CentraleSupélec move increasingly towards open science practices. Barriers are rather technical or expertise-based and can be eradicated with targeted measures if the willingness is there. Advancements in open data and methods across sectors have the potential to fast-track innovation and research and are crucial for increasingly data-driven policymaking and design methods.

This brings us to the end of this book. We hope to have been able to show that for the transition towards a sustainable and just future, plenty pathways and design approaches already exist today and many more are already—or wait to be—explored. More projects and research are needed that integrate *uncertainty* whilst being *transdisciplinary*, *people-centred*, *applied*, and *open*. We close by arguing that they can maximise the chances of success, as well as providing some hope and direction when facing the complex challenges of sustainable and people-centred urban mobility.

Glossary

Terms	*Acronyms*	*Description*
Accessibility		'Extent to which land use and transport systems enable (groups of) individuals to reach activities or destinations by means of a (combination of) transport mode(s)' (Geurs and van Wee, 2004, p. 128)
Active Mobility		Active mobility refers to all types of human-powered mobility, including that of walking, cycling, and using kick-scooters
Agent		Term to refer to a group of individuals/people in the context of predominantly agent-based simulations. A set of agents makes up a *synthetic population*
Agent-based simulation		Computational modelling technique used to simulate complex systems by representing individual agents and their interactions. Each agent is an autonomous entity with a set of rules, behaviours, and decision-making processes. Agents interact with each other and their environment, leading to emergent behaviour and patterns

(continued)

T. Gall et al., *Sustainable Urban Mobility Futures*, Sustainable Urban Futures, https://doi.org/10.1007/978-3-031-45795-1

(continued)

Terms	*Acronyms*	*Description*
Automated Vehicle	AV	Often referred to as a self-driving or autonomous vehicle. An automobile equipped with advanced technology and sensors that enable it to navigate and operate without human intervention. The term automated is chosen over autonomous as the latter implies fully autonomy which is considered far from current progress
Bus Rapid Transit	BRT	High-capacity bus service with dedicated lanes and stops as lower cost mass transit alternative compared to rail-based mass transit
Carbon dioxide equivalent	CO_2e	Standard unit used for combined effect of different GHGs in terms of their contribution to global warming. It provides a way to compare the warming potential of various gases based on their heat-trapping properties and their longevity in the atmosphere relative to carbon dioxide (CO_2). This includes primarily CO_2, methane (CH_4), and Nitrous Oxide (N_2O)
City		A socio-spatial construct that is defined by socially, economically, and spatially predominantly continuous human settlements. It is oftentimes set equal with the administrative boundary but might be larger or only make up a smaller part thereof (cf. Urban area)
Electric Vehicle	EV	Overall term for vehicles powered by electricity. It can be distinguished between battery-electric vehicle (BEV) and Plug-in Hybrid Electric Vehicle (PHEV), also commonly referred to as hybrid vehicles
FAIR principles	FAIR	Principles for (research) data management: Findable, Accessible, Interoperable, and Reusable
Future scenario		See 'scenario'

(continued)

(continued)

Terms	*Acronyms*	*Description*
Greenhouse gas emissions	GHG	More general term for emissions that contribute to the greenhouse effect, meaning the heating of the atmosphere due to higher reflection of radiation due to higher GHG concentration (cf. CO_2e)
Glocalisation		The parallel process of more global links whilst re-focusing on local connections and communities
General Transit Feed Specification	GTFS	Open standard for public transport, including information on elements such as routes, stops, schedules, or capacities
Intermodality		The combined and integrated use of several modes. E.g., a trip starting by taking a bike and parking it at a station, followed by taking a train
Low Emission Zone	LEZ	Geographic areas, either full *cities* or parts thereof, where emissions shall be reduced by restricting vehicles which are too emitting. "Such policy might permit only vehicles with emissions under a set standard (e.g., Euro emissions ratings) to enter a specified area or vehicle taxation depending on their emissions".
Multi-Agent-based Transport Simulation	MATSim	Open-source framework used for simulating and modelling urban mobility systems. It employs agent-based modelling to simulate the behaviour of individual agents as they make travel decisions within a transportation network (Horni et al., 2016)
Multimodality		The use of multiple modes from the same provider or within the same framework. For example, using a bus, metro, and tram on one trip, all operated by the same institution and accessed with the same ticket (Nobis, 2010)

(continued)

(continued)

Terms	*Acronyms*	*Description*
Mobility		Ability and ease of movement of people, goods, or information within and between different locations or points. Here the mobility of people across time and space to fulfil their daily life activities (Kaufmann, 2002, 2011)
Mobility-as-a-Service	MaaS	People-centred technological and organisational innovation aiming to improve mobility access and management by integrating data from different mobility-related services and providing users with real-time information, navigation, booking, and payment features (Heikkilä, 2014; Kamargianni et al., 2016; Jittrapirom et al., 2017; Sochor et al., 2018; Reyes Madrigal et al., 2023)
Variant		An option or an alternative of, e.g., a particular service. For example, we might design a metro system with a capacity of 20,000 people per hour per direction (p/h/d). A variant is a metro with 15,000 p/h/d or a light-rail system with 5,000 p/h/d. We do not consider scenarios as the adequate term to refer to these options or alternatives due to the confusion potential with future *scenarios*
Shared Automated Electric Vehicle	SAEV	Sub-set of *shared* automated vehicles. In case of larger vehicles, they are frequently referred to as shuttles. In the past, robotaxis has been used for the same meaning. Related concepts are those of Mobility-on-Demand (MoD) and in transport terminology, Demand-responsive Transport (DRT)
Scenario		Futures. Possible, plausible, and distinct future alternatives with a narrative description (Spaniol and Rowland, 2018). Not to be mixed up with *Variant*

(continued)

(continued)

Terms	*Acronyms*	*Description*
Sustainable Development Goal	SDG	'A universal call to action to end poverty, protect the planet, and ensure that by 2030 all people enjoy peace and prosperity' (UN, 2015). The 17 SDGs recognise interrelated areas of actions, searching for a balance between social, economic, and environmental sustainability
Sustainable Urban Mobility Plan	SUMP	Strategic planning approach used by cities and urban areas to develop and implement mobility plans that prioritise sustainable and efficient mobility options
Synthetic population		Synthetic populations are digital representations of the real population (Ramadan and Sisiopiku, 2020; Hermes and Poulsen, 2012). Synthetic populations represent reality in terms of individual persons, often grouped into households
People-centred design	PCD	Extension of terms for human- and used-centred designs with focus on consideration of the impacts of or on people across time and space (Gall et al., 2021)
Persona		Fictitious character representing a homogeneous class of users (Cooper, 1999). It is frequently used in user-centred design. A persona is traditionally represented by a name, a photo, and a narrative part describing attitudes and typical behaviours
Public Transport		Various modes of shared transportation services available to the general public, including buses, trains, trams, and ferries. The sub-category mass transit specifically focuses on large-capacity systems for efficiently moving many passengers within urban areas

(continued)

(continued)

Terms	*Acronyms*	*Description*
Transit-oriented Development		Urban planning strategy promoting the creation of compact, mixed-use communities centred around public transportation hubs, with focus on increased density (ITDP, 2020)
Transition		Transition from current to future state by addressing one or several problems of current state through specific solutions (Rotmans and Loorbach, 2009)
Trend		Anticipated future development with *high certainty*
Uncertainty(ies)		Anticipated future development with *high uncertainty*. Not to be confused with uncertainty which results from lack of precision or is the outcomes of computations or algorithm
Urban area		An area with urban core and surrounding functional urban area. Often not equal to administrative boundary (cf. OECD, 2019)
Urban form		Physical layout, structure, and arrangement of buildings, streets, open spaces, and infrastructure within a city or urban area. Includes land use patterns, building heights, densities, street networks, and public spaces (Oliveira, 2016)
Urban mobility system		Multiscale and multi-disciplinary complex adaptive socio-technical system that defines personal mobility taking place in the primarily continuous daily territories of the residents and users of an urban or metropolitan area (Gall, 2023)

(continued)

(continued)

Terms	*Acronyms*	*Description*
Urbanisation		Process where increasing proportion of a population migrates from rural areas to urban areas, leading to the growth and expansion of cities and towns
User		Customers and target groups for whom products and services are tailored for

References

Al Maghraoui, O., Vallet, F., Puchinger, J., and Yannou, B. (2019) Modeling traveler experience for designing urban mobility systems. *Design Science*, Vol. 5/E7. https://doi.org/10.1017/dsj.2019.6.

Agarwal, O.P., Zimmerman, S., and Kumar, A. (2018). *Emerging paradigms in urban mobility. Planning, financing and management*. Elsevier. ISBN: 9780128114353.

Aguilera, A., and Boutueil, V. (2019) *Urban mobility and the smartphone*. Elsevier. https://doi.org/10.1016/C2016-0-03595-7.

Al Maghraoui, O., Vallet, F., Puchinger, J., and Yannou, B. (2019) Modeling traveler experience for designing urban mobility systems. *Design Science*, Vol. 5/E7. https://doi.org/10.1017/dsj.2019.6.

Allam, Z., Bibri, S., Chabaud, D., and Moreno, C. (2022) The theoretical, practical, and technological foundations of the 15-minute city model: Proximity and its environmental, social and economic benefits for sustainability. *Energies*, Vol. 15. https://doi.org/10.3390/en15166042.

Amin, A., Tareen, W.U.K., Usman, M., Ali, H., Bari, I., Horan, B., Mekhilef, S., Asif, M., Ahmed, S., and Mahmood, A. (2020) A review of optimal charging strategy for electric vehicles under dynamic pricing schemes in the distribution charging network. *Sustainability*, Vol. 12, p. 10160. https://doi.org/10.3390/su122310160.

Anda, C., Ordonez Medina, S.A., and Axhausen, K.W. (2021) Synthesising digital twin travellers: Individual travel demand from aggregated mobile phone data. *Transportation Research Part C: Emerging Technologies*, Vol. 128, p. 103118. https://doi.org/10.1016/j.trc.2021.103118.

T. Gall et al., *Sustainable Urban Mobility Futures*, Sustainable Urban Futures, https://doi.org/10.1007/978-3-031-45795-1

Appleyard, D. (1982) *Livable streets*. San Francisco: University of California Press.

Arif, A.I., Babar, M., Imthias Ahamed, T.P., Al-Ammar, E.A., Nguyen, P.H., René Kamphuis, I.G., and Malik, N.H. (2016) Online scheduling of plug-in vehicles in dynamic pricing schemes. *Sustainable Energy, Grids and Networks*, Vol. 7, pp. 25–36. https://doi.org/10.1016/j.segan.2016.05.001.

Auld, J., Hope, M., Ley, H., Sokolov, V., Xu, B., and Zhang, K. (2016) POLARIS: Agent-based modeling framework development and implementation for integrated travel demand and network and operations simulations. *Transportation Research Part C: Emerging Technologies*, Vol. 64, pp. 101–116. https://doi.org/10.1016/j.trc.2015.07.017.

Autio, E., and Thomas, L. (2014) Innovation ecosystems. In: Dodgson, M., Gann, D.M., Phillips, N. (eds.), *The Oxford handbook of innovation management*. Oxford University Press, pp. 204–288.

Azevedo, C.L., Deshmukh, N.M., Marimuthu, B., Oh, S., Marczuk, K., Soh, H., Basak, K., Toledo, T., Peh, L.-S., and Ben-Akiva, M.E. (2017) Simmobility short-term: An integrated microscopic mobility simulator. *Transportation Research Record Journal of the Transportation Research Board*, Vol. 2622, pp. 13–23. https://doi.org/10.3141/2622-02.

Babiker, M., Bazaz, A., Bertoldi, P., Creutzig, F., De Coninck, H., De Kleijne, K., Dhakal, S., Haldar, S., Jiang, K., Kılkış, Ş., Klaus, I., Krishnaswamy, J., Lwasa, S., Niamir, L., Pathak, M., Pereira, J.P., Revi, A., Roy, J., Seto, K.C., Singh, C., Some, S., Steg, L., and Ürge-Vorsatz, D. (2022) *What the latest science on climate change mitigation means for cities and urban areas*. Indian Institute for Human Settlements. https://doi.org/10.24943/SUPSV310.2022.

Babst, V., Du Bernard, C., Gangloff, C., and Kautzmann, J.-E. (2022) Results based management. Approach of the Council of Europe. Practical Guide. Council of Europe. Directorate of Programme and Budget Programme division.

Balac, M., and Hörl, S. (2021) Synthetic population for the state of California based on open-data: examples of San Francisco Bay area and San Diego County. In 100th Annual Meeting of the Transportation Research Board. Washington, D.C., January 2021.

Balac, M., Ciari, F., and Axhausen, K.W. (2017) Modeling the impact of parking price policy on free-floating carsharing: Case study for Zurich, Switzerland. *Transportation Research Part C: Emerging Technologies*, Vol 77, pp. 207–225. https://doi.org/10.1016/j.trc.2017.01.022.

Ballard, B. (2007) *Designing the mobile user experience*, 1st ed. Wiley. https://doi.org/10.1002/9780470060575.

Baltazar, J., Vallet, and Garcia, J. (2022) A model for long-distance mobility with battery electric vehicles: A multi-perspective analysis. *Procedia CIRP*, Vol. 109, pp. 334–339. https://doi.org/10.1016/j.procir.2022.05.259.

Banister, D. (2008) The sustainable mobility paradigm. *Transport Policy*, Vol. 15/2, pp. 73–80. https://doi.org/10.1016/j.tranpol.2007.10.005.

Batty, M. (2018) Digital twins. *Environment and Planning B: Urban Analytics and City Science*, Vol. 45, pp. 817–820. https://doi.org/10.1177/2399808318796416.

Becker, H., Balac, M., Ciari, F., and Axhausen, K.W. (2020) Assessing the welfare impacts of Shared Mobility and Mobility as a Service (MaaS). *Transportation Research, Part A: Policy and Practice*, Vol. 131, pp. 228–243. https://doi.org/10.1016/j.tra.2019.09.027.

Bertolini, L. (2020) From "streets for traffic" to "streets for people": Can street experiments transform urban mobility? *Transport Reviews*, pp. 1–20. https://doi.org/10.1080/01441647.2020.1761907.

Bertram, B., and Berthold, B.F. (2012) Was Sind Sinus-Milieus®? In: Thomas, P.M. and Calmbach, M. (eds.), *Jugendliche Lebenswelten: Perspektiven für Politik, Pädagogik und Gesellschaft*. Berlin, Heidelberg: Springer Berlin Heidelberg, pp. 11–35.

Bertraud, A. (2018) *Order without design. How markets shape cities*. Cambridge/London: The MIT Press.

Bevan, L.D. (2022) The ambiguities of uncertainty: A review of uncertainty frameworks relevant to the assessment of environmental change. *Futures*, Vol. 137, p. 102919. https://doi.org/10.1016/j.futures.2022.102919.

Bishop, J.D., Axon, C.J., Bonilla, D., Tran, M., Banister, D., and McCulloch, M.D. (2013) Evaluating the impact of V2G services on the degradation of batteries in PHEV and EV. *Applied Energy*, Vol. 111, pp. 206–218.

Bongardt, D., Stiller, L., Swart, A., et al. (2019) Sustainable urban transport: Avoid-shift-improve (ASI).

Bonilla, R., and Carreon-Sosa, R. (2020). Transport and sustainability. In *International encyclopedia of human geography*, 2nd ed., Vol. 13. https://doi.org/10.1016/B978-0-08-102295-5.10150-7.

Bonnafous, A. (1996). Le système des transports urbains. *Économie et statistique*, Vol. 294/1, pp. 99–108. https://doi.org/10.3406/estat.1996.6087.

Börjeson, L., Höjer, M., Dreborg, K.-H., Ekvall, T., and Finnveden, G. (2006) Scenario types and techniques: Towards a user's guide. *Futures*, Vol. 38/7, pp. 723-739. https://doi.org/10.1016/j.futures.2005.12.002.

Bornet, C., and Brangier, É. (2013) La méthode des personas: Principes, intérêts et limites, *Bulletin de psychologie*, Vol. 524/2. https://doi.org/10.3917/bupsy.524.0115.

Borysov, S.S., Rich, J., and Pereira, F.C. (2019) How to generate micro-agents? A deep generative modeling approach to population synthesis. *Transportation*

Research Part C: Emerging Technologies, Vol. 106, pp. 73–97. https://doi.org/10.1016/j.trc.2019.07.006.

Bower, J.L., and Christensen, C.M. (1995) Disruptive innovation: Catching the wave. *Harvard Business Review*, Vol. 73/1, pp. 43–45.

Brög, W., Erl, E., and Mense, N. (2002) Individualised marketing changing travel behaviour for a better environment. http://www.epomm.eu/newsletter/v2/content/2015/1115/doc/IndiMark.pdf.

Brown, L. (2021) Hans Arby: MaaS—mars 2021. Futura-Mobility. https://futuramobility.org/fr/hans-arby-maas-mars-2021/ [accessed August 2023].

Bunn, M.D., Savage, G.T., and Holloway, B.B. (2002) Stakeholder analysis for multi-sector innovations. *Journal of Business & Industrial Marketing*, Vol. 17(2/3), pp. 181-203. https://doi.org/10.1108/08858620210419808.

Buur, J., and Matthews, B. (2008) Participatory innovation. *International Journal of Innovation Management*, Vol. 12/3, pp. 255–273. https://doi.org/10.1142/S1363919608001996.

Cao, C., and Chen, B. (2018) Generalized Nash equilibrium problem based electric vehicle charging management in distribution networks. *International Journal of Energy Research*, Vol. 42, pp. 4584–4596. https://doi.org/10.1002/er.4194.

Cao, T. et al. (2012) An optimized EV charging model considering TOU price and SOC curve. *IEEE Transactions on Smart Grid*, Vol. 3/1, pp. 388–393. https://doi.org/10.1109/TSG.2011.2159630.

CDC. (2020) Road traffic injuries and deaths—A global problem. Accessible at: https://www.cdc.gov [accessed August 2023].

CEBR. (2014) The future economic and environmental costs of gridlock in 2030. An assessment of the direct and indirect economic and environmental costs of idling in road traffic congestion to households in the UK, France, Germany and the USA. Centre for Economics and Business Research (CEBR).

Charreaux, G. (2004) Les théories de la gouvernance: de la gouvernance des entreprises à la gouvernance des systèmes nationaux (No. 1040101). Université de Bourgogne-CREGO EA7317 Centre de recherches en gestion des organisations.

Chen, Q. et al. (2017) Dynamic price vector formation model-based automatic demand response strategy for PV-assisted EV charging stations. *IEEE Transactions on Smart Grid*, Vol. 8/6, pp. 2903–2915. https://doi.org/10.1109/TSG.2017.2693121.

Chiş, A., Lundén, J., and Koivunen, V. (2016) Reinforcement learning-based plug-in electric vehicle charging with forecasted price. *IEEE Transactions on Vehicular Technology*, Vol. 66/5, pp. 3674–3684.

Chouaki, T., Reyes Madrigal, L.M., and Hörl, S. (2024) Assessing the impact of monetary incentives for walking using agent-based mobility simulations and discrete mode choice models. TRB Annual Meeting. Washington DC, USA, 07–11 January 2024 [forthcoming].

Chouaki, T. (2023). Agent-based simulations of intermodal mobility-on-demand systems operated by reinforcement learning. Doctoral dissertation, University Paris-Saclay.

Climate Watch. (2020) Historical GHG emissions. Accessible at: https://www.climatewatchdata.org [accessed August 2023].

Colantonio, A., and Dixon, T. (2009) Measuring socially sustainable urban regeneration in Europe, Oxford Institute for Sustainable Development (OISD), School of the Built Environment, Oxford Brookes University, 2009.

Cook, S., Stevenson, L., Aldred, R., Kendall, M., and Cohen, T. (2022) More than walking and cycling: What is 'active travel'? *Transport Policy*, Vol. 126, pp. 151–161. https://doi.org/10.1016/j.tranpol.2022.07.015.

Cooper, A. (1999) *The inmates are running the asylum*. New York: Macmillan.

COP21. (2015) Paris Agreement to the United Nations Framework Convention on Climate Change, December 12, 2015, T.I.A.S. No. 16-1104.

Courmont, A., and Le Galès, P. (2019) Gouverner la ville numérique. Ville et numérique, Presses Universitaires de France. 9782130815259.

Creutzig, F. (2016) Evolving narratives of low-carbon futures in transportation. *Transport Reviews*, Vol. 36/3, pp. 341–360. https://doi.org/10.1080/01441647.2015.1079277.

Dankl, K. (2017) Design age: Towards a participatory transformation of images of ageing. *Design Studies*, Vol. 48, pp. 30–42. https://doi.org/10.1016/j.destud.2016.10.004.

Dator, J. (2019) What futures studies is, and is not. In *Jim Dator: A noticer in time. Anticipation science*, Vol. 5. Cham: Springer. https://doi.org/10.1007/978-3-030-17387-6_1.

Dave, S. (2010) High urban densities developing countries: A sustainable solution? *Built Environment*, Vol. 36/1, pp. 9–27. https://www.jstor.org/stable/23289981.

De Bitencourt, L.A., Borba, B.S.M.C., Maciel, R.S., Fortes, M.Z., and Ferreira, V.H. (2017) Optimal EV charging and discharging control considering dynamic pricing. 2017 IEEE Manchester PowerTech, Manchester, UK, 2017, pp. 1–6. https://doi.org/10.1109/PTC.2017.7981231.

Debnath, A., Chin, H., Haque, M.M., and Yuen, B. (2014). A methodological framework for benchmarking smart transport cities. *Cities*, Vol. 37.

Diallo, A.O., Doniec, A., Lozenguez, G., and Mandiau, R. (2021) Agent-based simulation from anonymized data: An application to Lille metropolis. *Procedia Computer Science*, Vol. 184, pp. 164–171. https://doi.org/10.1016/j.procs.2021.03.027.

Dimitrov, S., and Lguensat, R. (2014) Reinforcement learning based algorithm for the maximization of EV charging station revenue. 2014 International

Conference on Mathematics and Computers in Sciences and in Industry. IEEE.

Docherty, I., Marsden, G., and Anable, J. (2018) The governance of smart mobility. *Transportation Research Part A: Policy and Practice*, Vol. 115, pp. 114–125. https://doi.org/10.1016/j.tra.2017.09.012.

Dong, L., Santi, P., Liu, Y., Zheng, S., and Ratti, C. (2022) The universality in urban commuting across and within cities. https://doi.org/10.48550/arXiv.2204.12865.

Dorokhova, M., Martinson, Y., Ballif, C., and Wyrsch, N. (2021) Deep reinforcement learning control of electric vehicle charging in the presence of photovoltaic generation. *Applied Energy*, Elsevier, Vol. 301(C). https://doi.org/10.1016/j.apenergy.2021.117504.

Duan, Y., Chen, X., Houthooft, R., Schulman, J., and Abbeel, P. (2016) Proceedings of the 33rd International Conference on Machine Learning, PMLR 48, pp. 1329–1338. https://doi.org/10.48550/arXiv.1604.06778.

Dubey, A., and Santoso, S. (2015) Electric vehicle charging on residential distribution systems: Impacts and mitigations. *IEEE Access*, Vol. 3, pp. 1871–1893. https://doi.org/10.1109/ACCESS.2015.2476996.

Durán-Heras, A., García-Gutiérrez, I., and Castilla-Alcalá, G. (2018) Comparison of iterative proportional fitting and simulated annealing as synthetic population generation techniques: Importance of the rounding method. *Computers Environment and Urban System*, Vol. 68, pp. 78–88. https://doi.org/10.1016/j.compenvurbsys.2017.11.001.

IDusparic, I., Harris, C., Marinescu, A., Cahill, V., and Clarke, S. (2013) Multi-agent residential demand response based on load forecasting. 2013 1st IEEE Conference on Technologies for Sustainability (SusTech), Portland, OR, USA, pp. 90–96. https://doi.org/10.1109/SusTech.2013.6617303.

Dyson, P., and Sutherland, R. (2021) *Transport for humans. Are we nearly there yet?* London Publishing Partnership.

European Commission (EC). (2011) Roadmap to a single European transport area—Towards a competitive and resource efficient transport system. EU white paper COM(2011) 144. Brussels: European Commission.

European Commission (EC). (2021) Delivering the European Green Deal. https://commission.europa.eu/strategy-and-policy/priorities-2019-2024/european-green-deal/delivering-european-green-deal [accessed August 2023].

Eendebak, R., and World Health Organization. (2015) *World report on ageing and health.* World Health Organization.

Elioth, Egis Group. (2017) Paris, an air of change. Towards carbon neutrality in 2050. https://paris2050.elioth.com/en/ [accessed August 2023].

Esmaili, M., and Goldoust, A. (2015) Multi-objective optimal charging of plug-in electric vehicles in unbalanced distribution networks. International *Journal of Electrical Power & Energy Systems*, Vol. 73, pp. 644–652. https://doi.org/10.1016/j.ijepes.2015.06.001.

Fergnani, A. (2019) The future persona: A futures method to let your scenarios come to life. *Foresight*. https://doi.org/10.1108/FS-10-2018-0086.

Fergnani, A., and Jackson, M. (2019) Extracting scenario archetypes: A quantitative text analysis of documents about the future, *Future & Foresight Science*, Vol. 1/2, pp. 1–14. https://doi.org/10.1002/ffo2.17.

Ferro, G., Laureri, F., Minciardi, R., and Robba, M. (2018) An optimization model for electrical vehicles scheduling in a smart grid. *Sustainable Energy, Grids and Networks*, Vol. 14, pp. 62–70. https://doi.org/10.1016/j.segan.2018.04.002.

Flipo, F., Dobré, M., and Michot, M. (2013) La face cachée du numérique . L'impact environnemental des nouvelles technologies. Le Kremlin Bicêtre: L'Échappée.

Fogg, B.J. (2009) A behavior model for persuasive design. In Proceedings of the 4th International Conference on Persuasive Technology (Persuasive'09). Association for Computing Machinery, New York, NY, USA, Article 40, pp. 1–7. https://doi.org/10.1145/1541948.1541999.

Fragkias, M., Lobo, J., Strumsky, D., and Seto, K.C. (2013) Does size matter? Scaling of CO_2 emissions and U.S. urban areas. *PloS One*, Vol. 8/6, pp. 1–8. https://doi.org/10.1371/journal.pone.0064727.

Fuglerud, K.S., Schulz, T., Janson, A.L., and Moen, A. (2020) Co-creating persona scenarios with diverse users enriching inclusive design. In: Antona M., Stephanidis C. (eds.), *Universal access in human-computer interaction*. Cham: Springer. https://doi.org/10.1007/978-3-030-49282-3_4.

Gall, T., and Haxhija, S. (2020) Storytelling of and for planning: Urban planning through participatory narrative-building, 56th ISOCARP World Planning Congress, Doha/online, 8 November 2020–4 February 2021.

Gall, T., Vallet, F., Douzou, S., & Yannou, B. (2021) Re-defining the system boundaries of human-centred design. *Proceedings of the Design Society*, Vol. 1, pp. 2521–2530. https://doi.org/10.1017/pds.2021.513.

Gall, T., Vallet, F., and Yannou, B. (2022) How to visualise futures studies concepts: Revision of the futures cone. *Futures*, Vol. 143, p. 103024. https://doi.org/10.1016/j.futures.2022.103024.

Geels, F.W. (2004). From sectoral systems of innovation to socio-technical systems: Insights about dynamics and change from sociology and institutional theory. *Research Policy*, Vol. 33/6, pp. 897–920. https://doi.org/10.1016/j.respol.2004.01.015.

Gehl, J., and Svarre, B. (2013) *How to study public life*. Washington, DC: Island Press.

Gehl, J. (2011) *Life between buildings: Using public space*, 6th ed. London: Island Press.

Glaeser, E. (ed.). (2010) *Agglomeration economies*. Chicago: The University of Chicago Press/National Bureau of Economic Research.

Gleave, S.D. (2016) Study on the prices and quality of rail passenger services (Final Report April 2016, 22870601/MOVE/B2/2015-126). European Commission Directorate General for Mobility and Transport. https://transport.ec.europa.eu/system/files/2016-09/2016-04-price-quality-rail-pax-services-resume-analytique.pdf.

Gössling, S., Ceron, J.-P., Dubois, G., and Hall, C.M. (2009) Hypermobile travellers. In: Gössling, S., and Upham, P. (eds.), *Climate change and aviation*. London: Earthscan.

Green, B. (2020) *The smart enough city. Putting technology in its place to reclaim our urban future*. Cambridge: The MIT Press.

Gregory, J. (2003) Scandinavian approaches to participatory design. *International Journal on Engineering Education*, Vol. 19/1, pp. 62–74.

Hanson, S. (2010) Gender and mobility: New approaches for informing sustainability. *Gender, Place & Culture*, Vol. 17/1, pp. 5–23. https://doi.org/10.1080/09663690903498225.

Hargreaves, T., Longhurst, N., and Seyfang, G. (2012) Understanding sustainability innovations: Points of intersection between the multi-level perspective and social practice theory. 3S Working Paper 2012-03. Norwich: Science, Society and Sustainability Research Group.

Heikkilä, S. (2014) Mobility as a service—A proposal for action for the public administration, case Helsinki. http://urn.fi/URN:NBN:fi:aalto-201405221895.

Hensher, D.A., and Hietanen, S. (2023) Mobility as a feature (MaaF): Rethinking the focus of the second generation of mobility as a service (MaaS). *Transport Reviews*, Vol. 43/3, pp. 325–329. https://doi.org/10.1080/01441647.2022.2159122.

Héran, F., and Ravalet, E. (2008) La consummation d'espace-temps des divers modes de déplacement en milieu urbain: Application au cas de l'Ile de France. Lettre de commande 06 MT E012. Ministère des transports, de l'équipement, du tourisme et de la mer.

Hermes, K., and Poulsen, M. (2012) A review of current methods to generate synthetic spatial microdata using reweighting and future directions. *Computers, Environment and Urban Systems*, Vol. 36/4, pp. 281–290. https://doi.org/10.1016/j.compenvurbsys.2012.03.005.

Hietanen, S. (2014) 'Mobility as a service'—The new transport model? *Eurotransport*, Vol. 12/2. ITS & Transport Management Supplement, pp. 2–4.

Hillier, B. (2009) Spatial sustainability in cities: Organic patterns and sustainable forms. Proceedings of the 7th International Space Syntax Symposium. Stockholm: KTH.

Hirschhorn, F., Paulsson, A., Sørensen, C.H., and Veeneman, W. (2019) Public transport regimes and mobility as a service: Governance approaches in Amsterdam, Birmingham, and Helsinki. *Transportation Research Part A: Policy and Practice*, Vol. 130, pp. 178–191. https://doi.org/10.1016/j.tra.2019.09.016.

Ho et al. (2018) Potential uptake and willingness-to-pay for Mobility as a Service (MaaS): A stated choice study. *Transportation Research Part A*, Vol. 117, pp. 302–318. https://doi.org/10.1016/j.tra.2018.08.025.

Hoornweg, D., and Pope, K. (2017) Population predictions for the world's largest cities in the 21st century. *Environment and Urbanization*, Vol. 29/1, pp. 195–216. https://doi.org/10.1177/0956247816663557.

Hörl, S., and Axhausen, K.W. (2021) Relaxation–discretization algorithm for spatially constrained secondary location assignment. *Transportmetrica A: Transport Science*, Vol. 19/2, pp. 1–20. https://doi.org/10.1080/23249935.2021.1982068.

Hörl, S., and Balac, M. (2021a) Synthetic population and travel demand for Paris and Île-de-France based on open and publicly available data. *Transportation Research Part C: Emerging Technologies*, Vol. 130, p. 103291. https://doi.org/10.1016/j.trc.2021.103291.

Hörl, S., and Balac, M. (2021b) Open synthetic travel demand for Paris and Île-de-France: Inputs and output data. *Data Brief*, Vol. 39, p. 107622. https://doi.org/10.1016/j.dib.2021.107622.

Hörl, S., and Puchinger, J. (2022) From synthetic population to parcel demand: Modeling pipeline and case study for last-mile deliveries in Lyon. Presented at the Transport Research Arena (TRA) 2022, Lisbon.

Hörl, S., Ruch, C., Becker, F., Frazzoli, E., and Axhausen, K.W. (2019) Fleet operational policies for automated mobility: A simulation assessment for Zurich. *Transportation Research Part C: Emerging Technologies*, Vol. 102, pp. 20–31. https://doi.org/10.1016/j.trc.2019.02.020.

Hörl, S., Becker, F., and Axhausen, K.W. (2021) Simulation of price, customer behaviour and system impact for a cost-covering automated taxi system in Zurich. *Transportation Research Part C: Emerging Technologies*, Vol. 123, p. 102974. https://doi.org/10.1016/j.trc.2021.102974.

Hörl, S. (2023) Towards replicable mode choice models for transport simulations in France. *9th International Symposium on Transportation Data & Modelling (ISTDM2023)*, 19-22 June 2023, Ispra.

Horni, A., Nagel, K., and Axhausen, K.W. (eds.). (2016). *The multi-agent transport simulation MATSim*. London: Ubiquity Press. https://doi.org/10.5334/baw.

Horschutz Nemoto, E., Issaoui, R., Korbee, D., Jaroudi, I., and Fournier, G. (2021) How to measure the impacts of shared automated electric vehicles on urban mobility. *Transportation Research Part D: Transport and Environment*, Vol. 93, p. 102766. https://doi.org/10.1016/j.trd.2021.102766.

Hu, J., Saleem, A., You, S., Nordström, L., Lind, M., and Østergaard, J. (2015) A multi-agent system for distribution grid congestion management with electric vehicles. *Engineering Applications of Artificial Intelligence*, Vol. 38, pp. 45–58. https://doi.org/10.1016/j.engappai.2014.10.017.

Huré, M. (2019) *Les mobilités partagées. Régulation politique et capitalisme urbain*. Paris: Editions de la Sorbonne.

IDEO. (2015) *A field guide to human-centred design*. 1st ed.

IEA. (2023) *Global EV Outlook 2023*. Paris: IEA.

Insee. (2022) Causes de décès en 2017: Comparaisons régionales et départementales [Gov]. Institut National de La Statistique et Des Études Économiques (Insee). https://www.insee.fr/fr/statistiques/2012788.

IPCC. (2022) Climate change 2022: Impacts, adaptation, and vulnerability. Contribution of Working Group II to the Sixth Assessment Report of the Intergovernmental Panel on Climate Change [H.-O. Pörtner, D.C. Roberts, M. Tignor, E.S. Poloczanska, K. Mintenbeck, A. Alegría, M. Craig, S. Langsdorf, S. Löschke, V. Möller, A. Okem, B. Rama (eds.)]. Cambridge/New York: Cambridge University Press. https://doi.org/10.1017/9781009325844.

ITDP. (2020) *What is TOD?* Institute for Transportation & Development Policy (ITDP). https://www.itdp.org/library/standards-and-guides/tod3-0/what-is-tod/.

ITF. (2021) Travel transitions: How transport planners and policy makers can respond to shifting mobility trends. ITF Research Reports. Paris: OECD Publishing.

Jabareen, Y.R. (2006) Sustainable urban forms: Their typologies, models, and concepts. *Journal of Planning Education and Research*, Vol. 26, pp. 38–52. https://doi.org/10.1177/2F0739456X05285119.

Jacobs, J. (1961) *The death and life of Great American cities*. New York (NY): Random House.

Jittrapirom, P., Caiati, V., Feneri, A., Ebrahimigharehbaghi, S., González, M., and Narayan, J. (2017) Mobility as a service: A critical review of definitions, assessments of schemes, and key challenges. *Urban Planning*, Vol. 2/2, pp. 13–25. https://doi.org/10.17645/up.v2i2.931.

Joubert, J.W., and de Waal, A. (2020) Activity-based travel demand generation using Bayesian networks. *Transportation Research Part C: Emerging Technologies*, Vol. 120, p. 102804. https://doi.org/10.1016/j.trc.2020.102804.

Jouffe, Y. (2014) La mobilité des pauvres. Contraintes et tactiques. *Informations sociales*, Vol. 182, pp. 90–99. https://doi.org/10.3917/inso.182.0090.

Kaack, L.H., Donti, P.L., Strubell, E., Kamiya, G., Creutzig, F., and Rolnick, D. (2022) Aligning artificial intelligence with climate change mitigation. *Nature*

Climate Change, Vol. 12, pp. 518–527. https://doi.org/10.1038/s41558-022-01377-7.

Kaddoura, I., Bischoff, J., and Nagel, K. (2020) Towards welfare optimal operation of innovative mobility concepts: External cost pricing in a world of shared autonomous vehicles. *Transportation Research Part A: Policy and Practice*, Vol. 136, pp. 48–63. https://doi.org/10.1016/j.tra.2020.03.032.

Kamargianni, M., Li, W., Matyas, M., and Schäfer, A. (2016) A critical review of new mobility services for urban transport. *Transport Research Arena, TRA2016*, Vol. 14, pp. 3294–3303.

L'Institut Paris Region. (2019) *Cities change the world*. Paris: L'Institut Paris Region.

Karfopoulos, E.L. and Hatziargyriou, N.D. (2012) A Multi-Agent System for Controlled Charging of a Large Population of Electric Vehicles. *IEEE Transactions on Power Systems*, Vol. 28/2, pp. 1196–1204. https://doi.org/10.1109/TPWRS.2012.2211624/

Katzmarzyk, P.T., Friedenreich, C., Shiroma, E.J., and Lee, I.M. (2022) Physical inactivity and non-communicable disease burden in low-income, middle-income and high-income countries. *British Journal of Sports Medicine*, Vol. 56/2, pp. 101–106. https://doi.org/10.1136/bjsports-2020-103640.

Kaufmann, V. (2002) *Re-thinking mobility. Contemporary sociology*. Hampshire: Ashgate.

Kaufmann, V. (2011) *Les Paradoxes de la Mobilité. Bouger, s'enraciner*, 2nd ed. Lausanne: Collection Le Savoir Suisse.

Khisty, C.J., and Zeitler, U. (2001) Is hypermobility a challenge for transport ethics and systemicity? *Systemic Practice and Action Research*, Vol. 14, pp. 597–613. https://doi.org/10.1023/A:1011925203641.

Knapskog, M., Hagen, O.H., Tennøy, A., and Rynning, M.K. (2019) Exploring ways of measuring walkability. *Transportation Research Procedia*, Vol. 41, pp. 264–282. https://doi.org/10.1016/j.trpro.2019.09.047.

König, D., Eckhardt, J., Aapaoja, A., Sochor, J., and Karlsson, M. (2016) Deliverable 3: Business and operator models for MaaS. MAASiFiE project funded by CEDR.

Korolko, N., and Sahinoglu, Z. (2015) Robust optimization of EV charging schedules in unregulated electricity markets. *IEEE Transactions on Smart Grid*, Vol. 8/1, pp. 149–157. https://doi.org/10.1109/TSG.2015.2472597.

Kropotkin, P. (1902) *Mutual aid: A factor of evolution*. London: Freedom Press.

Lajas, R., and Macário, R. (2020). Public policy framework supporting "mobility-as-a-service" implementation. *Research in Transportation Economics*, Vol. 83, p. 100905. https://doi.org/10.1016/j.retrec.2020.100905.

Lascoumes, P., and Le Galès, P. (2005). *Gouverner par les instruments* (Vol. 200). Paris: Presses de Sciences Po.

Law, R. (1999) Beyond 'women and transport': Towards new geographies of gender and daily mobility. *Progress in Human Geography*, Vol. 23/4, pp. 567–588.

Le Bescond, V., Can, A., Aumond, P., and Gastineau, P. (2021) Open-source modeling chain for the dynamic assessment of road traffic noise exposure. *Transportation Research Part D: Transport and Environment*, Vol. 94, p. 102793. https://doi.org/10.1016/j.trd.2021.102793.

Le Breton, E. (2019). *Mobilité, la fin du rêve?* Paris: Éditions Apogée.

Le Néchet, F. (2012) Urban spatial structure, daily mobility and energy consumption: A study of 34 European cities. *Cybergeo: European Journal of Geography, Sistemas, Modelística, Geoestadísticas*, Vol. 580. https://doi.org/10.4000/cybergeo.24966.

Leblond, V., Desbureaux, L., and Bielecki, V. (2020) A new agent-based software for designing and optimizing emerging mobility services: Application to city of Rennes. In European Transport Conference 2020.

Lee, U., Kang, N., and Lee, I. (2020) Shared autonomous electric vehicle design and operations under uncertainties: A reliability-based design optimization approach. *Structural Multidisciplinary Optimization*, Vol. 61, pp. 1529–1545.

Lefèvre, B., and Mainguy, G. (2009) Urban transport energy consumption: Determinants and strategies for its reduction, S.A.P.I.EN.S. *Surveys and Perspectives Integrating Environment and Society*, Vol. 2/3, pp. 1–18.

Leng, N., and Corman, F. (2020) The role of information availability to passengers in public transport disruptions: An agent-based simulation approach. *Transportation Research Part A: Policy and Practice*, Vol. 133, pp. 214–236. https://doi.org/10.1016/j.tra.2020.01.007.

Lesteven, G. et al. (2018). La transformation numérique des mobilités. Le nouveau monde de la mobilité, Presses des Ponts, pp. 141–146.

Li, H., Wan, Z., and He, H. (2019) Constrained EV charging scheduling based on safe deep reinforcement learning. *IEEE Transactions on Smart Grid*, Vol. 11/3, pp. 2427–2439. https://doi.org/10.1109/TSG.2019.2955437.

Liu, R., Dow, L., and Liu, E. (2011) A survey of PEV impacts on electric utilities. ISGT 2011, Anaheim, pp. 1–8. https://doi.org/10.1109/ISGT.2011.5759171.

Lopez, P.A., Wiessner, E., Behrisch, M., Bieker-Walz, L., Erdmann, J., Flotterod, Y.-P., Hilbrich, R., Lucken, L., Rummel, J., and Wagner, P. (2018) Microscopic traffic simulation using SUMO, in: 2018 21st International Conference on Intelligent Transportation Systems (ITSC). Presented at the 2018 21st International Conference on Intelligent Transportation Systems (ITSC), IEEE, Maui, HI, pp. 2575–2582. https://doi.org/10.1109/ITSC.2018.8569938.

Louf, R., and Barthelemy, M. (2014) Scaling: Lost in smog. *Environment and Planning B: Planning and Design*, Vol. 41, pp. 767–769. https://doi.org/10.1068/2Fb4105c.

L'usine nouvelle. (2018) Mobike devient le quatrième opérateur de vélos en libre-service dans Paris. https://www.usinenouvelle.com/article/mobike-devient-le-quatrieme-operateur-de-velos-en-libre-service-dans-paris.N642938 [accessed August 2023].

Lynch, P., and Horton, S. (2016) *Web style guide: Foundations of user experience design*, 4th ed. Yale University Press.

Maciejewski, M., Bischoff, J., Hörl, S., and Nagel, K. (2017) Towards a testbed for dynamic vehicle routing algorithms. In: Bajo, J., Vale, Z., Hallenborg, K., Rocha, A.P., Mathieu, P., Pawlewski, P., Del Val, E., Novais, P., Lopes, F., Duque Méndez, N.D., Julián, V., and Holmgren, J. (eds.), *Highlights of practical applications of cyber-physical multi-agent systems, communications in computer and information science.* Cham: Springer International Publishing. https://doi.org/10.1007/978-3-319-60285-1_6.

Manley, S. (2011). Chapter 17. Creating an accessible public realm. In: Smith, K.H., and Preiser, W.F.E. (eds.), *Universal design handbook*, 2nd ed. McGraw-Hill, pp. 69–79.

Manser, P., Becker, H., Hörl, S., and Axhausen, K.W. (2020) Designing a large-scale public transport network using agent-based microsimulation. *Transportation Research Part A: Policy and Practice*, Vol 137, pp. 1–15. https://doi.org/10.1016/j.tra.2020.04.011.

Marchetti, C. (1994) Anthropological invariants in travel behavior. *Technological Forecasting and Social Change*, Vol. 47/1.

Metabolic. (2019) *Metal demand for electric vehicle: Recommendations for fair, resilient, and circular transport systems*. Amsterdam: Metabolic.

Miaskiewicz, T. and Kozar, K. A. (2011) Personas and user-centered design: How can personas benefit product design processes? *Design Studies*, Vol. 32/5, pp. 417–430. https://doi.org/10.1016/j.destud.2011.03.003.

Miskolczi, M., Földes, D., Munkácsy, A., and Jászberényi, M. (2021) Urban mobility scenarios until the 2030s. *Sustainable Cities and Society*, Vol. 72, p. 103029. https://doi.org/10.1016/j.scs.2021.103029.

Mitchell, R.K., Agle, B.R., and Wood, D.J. (1997) Toward a theory of stakeholder identification and salience: Defining the principle of who and what really counts. *The Academy of Management Review*, Vol. 22/4, pp. 853–886. https://doi.org/10.2307/259247.

Mnih, V., Kavukcuoglu, K., Silver, D. et al. (2015) Human-level control through deep reinforcement learning. *Nature*, Vol. 518, pp. 529–533. https://doi.org/10.1038/nature14236.

Montgomery, C. (2013). *Happy city: Transforming our lives through urban design*. New York: Farrar, Straus and Giroux.

Mouly-Aigrot, B., Fouco, L., Leurent, F., and Lesteven, G. (2016) La transformation numérique nouvel eldorado pour les acteurs des transports?

Müller, K. (2017) *A generalized approach to population synthesis*. ETH Zurich. https://doi.org/10.3929/ETHZ-B-000171586.

Mulley, C. (2017) Mobility as a Services (MaaS)—Does it have critical mass? *Transport Reviews*, Vol. 37/3, pp. 247–251. https://doi.org/10.1080/01441647.2017.1280932.

Namazi-Rad, M.-R., Tanton, R., Steel, D., Mokhtarian, P., and Das, S. (2017) An unconstrained statistical matching algorithm for combining individual and household level geo-specific census and survey data. *Computers, Environment and Urban Systems*, Vol. 63, pp. 3–14. https://doi.org/10.1016/j.compenvurbsys.2016.11.003.

National Academies of Sciences, Engineering, and Medicine. (2016) *Between public and private mobility: Examining the rise of technology-enabled transportation services*. Washington, DC: The National Academies Press. https://doi.org/10.17226/21875.

Newman, P., and Kenworthy, J. (1989) *Cities and automobile dependence: An international sourcebook*. Aldershot: Gower.

Newsham, G.R., and Bowker, B.G. (2010) The effect of utility time-varying pricing and load control strategies on residential summer peak electricity use: A review. *Energy Policy*, Vol. 38/7, pp. 3289–3296. https://doi.org/10.1016/j.enpol.2010.01.027.

Nussbaum, M. (2003) Capabilities as fundamental entitlements: Sen and social justice. *Feminist Economic*, Vol. 9/2–3, pp. 33–59. https://doi.org/10.1080/1354570022000077926.

OECD. (1995) Environmental principles and concepts (OCDE/GD(95)124). Organisation for Economic Co-operation and Development (OECD). https://one.oecd.org/document/OCDE/GD(95)124/En/pdf.

OECD. (2019) Functional Urban Areas France, Regional Statistics, January 2019.

Oliveira, E.A., Andrade, J.S., and Makse, H.A. (2014) Large cities are less green. *Scientific Reports*, Vol. 4, pp. 1–12. https://doi.org/10.1038/srep04235.

Oliveira, V. (2016) *Urban morphology: An introduction to the study of the physical form of cities*. The Urban Book Series. Basel: Springer International Publishing.

Paap, J., and Katz, R. (2004). Anticipating disruptive innovation. *Research-Technology Management*, Vol. 47/5, pp. 13–22.

Pangbourne, K., Stead, D., Mladenović, M., and Milakis, D. (2018) The case of mobility as a service: A critical reflection on challenges for urban transport and mobility governance. *Governance of the Smart Mobility Transition*, pp. 33–48.

Piattoni, S. (2010) *The theory of multi-level governance: Conceptual, empirical, and normative challenges*. Oxford: Oxford University Press. https://doi.org/10.1093/acprof:oso/9780199562923.001.0001.

Picon, A. (2015) *Smart cities: A spatialised intelligence*. Chichester: Wiley.

Pont, M.B., and Haupt, P. (2009) Space, density and urban form. Doctoral Thesis, Delft: Technical University Delft.

Pruitt, J., and Grudin, J. (2003) Personas: Practice and theory, conference on designing for user experiences, San Francisco, June 2003, Association for Computing Machinery, New York (NY), pp. 1–15. https://doi.org/10.1145/997078.997089.

Qian, T., Shao, C., Wang, X., and Shahidehpour, M. (2020) Deep reinforcement learning for EV charging navigation by coordinating smart grid and intelligent transportation system. *IEEE Transactions on Smart Grid*, Vol. 11/2, pp. 1714–1723. https://doi.org/10.1109/TSG.2019.2942593.

Qian, T., Shao, C., Li, X., Wang, X., Chen, Z., and Shahidehpour, M. (2022) Multi-agent deep reinforcement learning method for EV charging station game. *IEEE Transactions on Power Systems*, Vol. 37/3, pp. 1682–1694. https://doi.org/10.1109/TPWRS.2021.3111014.

Ramadan, O.E., and Sisiopiku, V.P. (2020) A critical review on population synthesis for activity- and agent-based transportation models. In: De Luca, S., Di Pace, R., Djordjevic, B. (eds.), *Transportation systems analysis and assessment*. IntechOpen. https://doi.org/10.5772/intechopen.86307.

Randall, T. (2016) Here's how electric cars will cause the next oil crisis. A shift is under way that will lead to widespread adoption of EVs in the next decade. Bloomberg New Energy Finance 25.

Reckwitz, A. (2002) Toward a theory of social practices: A development in culturalist theorizing. *European Journal of Social Theory*, Vol. 5/2, pp. 243–263. https://doi.org/10.1177/13684310222225432.

Reed, M.S., Graves, A., Dandy, N., Posthumus, H., Hubacek, K., Morris, J., Prell, C., Quinn, C.H., and Stringer, L.C. (2009) Who's in and why? A typology of stakeholder analysis methods for natural resource management. *Journal of Environmental Management*, Vol. 90/5, pp. 1933–1949. https://doi.org/10.1016/j.jenvman.2009.01.001.

Reyes Madrigal, L.M. (2020) Mobility as a Service (MaaS): Configurations de gouvernance comme éléments de consolidation ou fragilisation des services de transport public (Mémoire de master 2e année: Urbanisme et aménagement. Transport et Mobilité). Université Gustave Eiffel.

Reyes Madrigal, L.M. (2024) Walking in Mobility as a Service (MaaS): Action levers for enhancing the integration of walking in MaaS from a stakeholder's ecosystem approach. Doctoral Dissertation, University Paris-Saclay.

Reyes Madrigal, L.M., Nicolaï, I., and Puchinger, J. (2023) Pedestrian mobility in Mobility as a Service (MaaS): Sustainable value potential and policy implications in the Paris region case. *European Transport Research Review*, Vol. 15/13. https://doi.org/10.1186/s12544-023-00585-2.

Rifkin, J. (2011) *The third industrial revolution*. London: Palgrave Macmillan.

Rittel, H.W.J., and Webber, M.M. (1973) Dilemmas in a general theory of planning. *Policy Sciences*, Vol. 4, pp. 155–169.

Robomobile Life Workshop. (2020) Prospective Atlas of the Robomobile Planet. The robomobile life. Prospective workshop. Paris: La Vie Robomobile.

Rohr, C., Ecola, L., Zmud, J., Dunkerly, F., Black, J., and Baker, E. (2016) *Travel in Britain in 2035: Future scenarios and their implications for technology innovation*. Santa Monica/Cambridge: RAND Corporation.

Rothenberg, S. (2007) Sustainability through servicizing. *MIT Sloan Management Review*, Vol. 48/2.

Rotmans, J., and Loorbach, D. (2009) Complexity and transition management. *Journal of Industrial Ecology*, Vol. 13/2, pp. 184–196. https://doi.org/10.1111/j.1530-9290.2009.00116.x.

Roy, W., and Yvrande-Billon, A. (2007). Ownership, contractual practices and technical efficiency: The case of urban public transport in France. *Journal of Transport Economics and Policy*, Vol. 41/2, pp. 257–282.

Ruelens, F., et al. (2012) Demand side management of electric vehicles with uncertainty on arrival and departure times. 2012 3rd IEEE PES Innovative Smart Grid Technologies Europe (ISGT Europe), Berlin, Germany, pp. 1–8. https://doi.org/10.1109/ISGTEurope.2012.6465695.

Saadi, I., Mustafa, A., Teller, J., Farooq, B., and Cools, M. (2016) Hidden Markov Model-based population synthesis. *Transportation Research Part B: Methodological*, Vol. 90, pp. 1–21. https://doi.org/10.1016/j.trb.2016.04.007.

Salet et al. (2003) *Metropolitan governance and spatial planning. Comparative case studies of European city-regions*. Spon-Press. Taylor and Francis group.

Sallard, A., Balać, M., and Hörl, S. (2021) An open data-driven approach for travel demand synthesis: An application to São Paulo. *Regional Studies, Regional Science*, Vol. 8, pp. 371–386. https://doi.org/10.1080/21681376.2021.1968941.

Sallis, J.F., and Owen, N. (1998) *Physical activity and behavioral medicine*. SAGE Publications.

Salminen, J., Santos, J.M., Jung, S.-G., Eslami, M., and Jansen, B.J. (2020) Persona transparency: Analyzing the impact of explanations on perceptions of data-driven personas. *International Journal of Human-Computer Interaction*, Vol. 36/8, pp. 788–800. https://doi.org/10.1080/10447318.2019.1688946.

Sanders, E.B.-N., and Stappers, P.J. (2008) Co-creation and the new landscapes of design. *International Journal of Co Creation in Design and the Arts*, Vol. 4, pp. 5–18. https://doi.org/10.1080/15710880701875068.
Sanders, E. B.-N., and Stappers, P.J. (2014) Probes, toolkits and prototypes: Three approaches to making in codesigning. *CoDesign*, Vol. 10/1, pp. 5–14, https://doi.org/https://doi.org/10.1080/15710882.2014.888183.
Sarasini, S., Sochor, J., and Arby, H. (2017) What characterises a sustainable MaaS business model? ICoMaaS 2017 Proceedings, 2017.
Schäfer, K., Rasche, P., Bröhl, C., Theis, S., Barton, L., Brandl, C., Wille, M, Nitsch, V., and Mertens, A. (2019) Survey-based personas for a target-group-specific consideration of elderly end users of information and communication systems in the German health-care sector. *International Journal of Medical Informatics*, Vol. 132, p. 103924. https://doi.org/10.1016/j.ijmedinf.2019.07.003.
Schmeer, K. (1999) Stakeholder analysis guidelines. https://dev2.cnxus.org/wp-content/uploads/2022/04/Stakeholders_analysis_guidelines.pdf [accessed August 2023].
Sen, A. (1979) *Equality of what? The Tanner lecture on human values*. Stanford University, 22 May 1979.
Seshadri, P., Joslyn, C., Hynes, M., and Reid, T. (2019) Compassionate design: Considerations that impact the users' dignity, empowerment and sense of security. *Design Science*, Vol. 5/E21. https://doi.org/10.1017/dsj.2019.18.
Seyfang, G., and Smith, A. (2007) Grassroots innovations for sustainable development: Towards a new research and policy agenda. *Environmental Politics*, Vol. 16/4, pp. 584–603, https://doi.org/ https://doi.org/10.1080/09644010701419121.
Sheller, M. (2013) Sociology After the Mobilities Turn from: The Routledge Handbook of Mobilities Routledge.
Sheller, M. (2018) Theorising mobility justice. *Tempo Social, revista de sociologia da USP*, Vol. 30/2, pp. 17–34.
Sheller, M., and Urry, J. (2006) The new mobilities paradigm. *Environment and Planning*, Vol. 38/2, pp. 207–226. https://doi.org/10.1068/a37268.
Silva, F.L.D., Nishida, C.E.H., Roijers, D.M., and Costa, A.H.R. (2020) Coordination of electric vehicle charging through multiagent reinforcement learning. *IEEE Transactions on Smart Grid*, Vol. 11/3, pp. 2347–2356. https://doi.org/10.1109/TSG.2019.2952331.
Singh, V.P., Kishor, N., and Samuel, P. (2017) Distributed multi-agent system-based load frequency control for multi-area power system in smart grid. *IEEE Transactions on Industrial Electronics,* Vol. 64/6, pp. 5151–5160.
Smith, G. (2020) Making mobility-as-a-service: Towards governance principles and pathways. Doctoral Dissertation. Chalmers University of Technology.

Soares, J., et al. (2017) Dynamic electricity pricing for electric vehicles using stochastic programming, Vol. 122, pp. 111–127.

Sochor, J., Arby, H., Karlsson, I.C.M., and Sarasini, S. (2018) A topological approach to mobility as a service: A proposed tool for understanding requirements and effects, and for aiding the integration of societal goals. *Research in Transportation Business & Management*, Vol. 27, pp. 3–14. https://doi.org/10.1016/j.rtbm.2018.12.003.

Spaniol, M.J., and Rowland, N.J. (2018) Defining scenario. *Futures & Foresight Science*, Vol 1/1. https://doi.org/10.1002/ffo2.3.

Spickermann, A., Grienitz, V., and von der Gracht, H.A. (2014) Heading towards a multimodal city of the future? *Technological Forecasting and Social Change*, Vol. 89(C), pp. 201–221.

Stead, D. (2016) Identifying key research themes for sustainable urban mobility. *International Journal of Sustainable Transportation*, 8318(February), Vol. 10/1, pp. 1–8.

Stevenson, P.D., and Mattson, C.A. (2019) The personification of big data. International Conference on Engineering Design ICED19, Delft, August 2019. https://doi.org/10.1017/dsi.2019.409.

Story, M.F. (2011) Chapter 4. The principles of universal design. In: Smith, K.H., and Preiser, W.F.E. (eds.), *Universal design handbook*, 2nd ed.. McGraw-Hill.

Streeting, M., Chen, H., and Edgar, E. (2017) Mobility as a service: The next transport disruption. Special Report. LEK Consulting. International Association of Public Transport (UITP) Australia & New Zealand and Tourism & Transport Forum (TTF) Australia.

Sun, L., and Erath, A. (2015) A Bayesian network approach for population synthesis. *Transportation Research Part C: Emerging Technologies*, Vol. 61, pp. 49–62. https://doi.org/10.1016/j.trc.2015.10.010.

Suyono, H., Rahman, M.T., Mokhlis, H., Othman, M., Illias, H.A., and Mohamad, H. (2019) Optimal scheduling of plug-in electric vehicle charging including time-of-use tariff to minimize cost and system stress. *Energies*, Vol. 12/8. https://doi.org/10.3390/en12081500.

Taylor, B.D. (2017) The geography of urban transportation finance. In: Giuliano, G., and Hanson, S. (eds.), *The geography of urban transportation*, 4th ed. New York/London: The Guilford Press.

Thomas, L.D.W., and Ritala, P. (2022) Ecosystem legitimacy emergence: A collective action view. *Journal of Management*, Vol. 48/3, pp. 515–541. https://doi.org/10.1177/0149206320986617.

Townsend, A. (2014) *Re-programming mobility: The digital transformation of transportation in the United States*. New York: The NYU Wagner Rudin Center.

Train, K. (2009) *Discrete choice methods with simulation*, 2th ed. Cambridge/ New York: Cambridge University Press.

Treib, O., Bähr, H., and Falkner, G. (2007) Modes of governance: Towards a conceptual clarification. *Journal of European Public Policy*, Vol. 14/1, pp. 1–20. https://doi.org/10.1080/13501760601071406.

Tuchnitz, F., Ebell, N., Schlund, J., and Pruckner, M. (2021) Development and evaluation of a smart charging strategy for an electric vehicle fleet based on reinforcement learning. *Applied Energy*, Vol. 285, p. 116382. https://doi.org/10.1016/j.apenergy.2020.116382.

UN. (2015) Draft outcome document of the United Nations summit for the adoption of the post-2015 development agenda. Transforming our world: the 2030 Agenda for Sustainable Development.

UNEP. (2008) Urban density and transport-related energy consumption. UNEP/GRID-Arendal Maps and Graphics Library.

UN-Habitat. (2016) *New urban agenda*. Nairobi: UN-Habitat.

UN-Habitat. (2022) Envisaging the future of cities. World Cities Report 2022. Nairobi: UN-Habitat.

University of Cambridge. (2023) *Inclusive design toolkit*. University of Cambridge. http://www.inclusivedesigntoolkit.com/ [accessed August 2023].

Urry, J. (2002) Mobility and proximity. *Sociology*, Vol. 36/2, pp. 255–274. https://doi.org/10.1177/0038038502036002002.

Urry, J. (2007) *Mobilities*. Hoboken: Wiley.

Urry, J. (2016) *What is the future?* Cambridge: Policy Press.

Usman, M., Tareen, W.U.K., Amin, A., Ali, H., Bari, I., Sajid, M., Seyedmahmoudian, M., Stojcevski, A., Mahmood, A., and Mekhilef, S. (2021) A coordinated charging scheduling of electric vehicles considering optimal charging time for network power loss minimization. *Energies*, Vol. 14/17. https://doi.org/10.3390/en14175336.

Vallet, F., Puchinger, J., Millonig, Al., and Lamé, G. (2020) Tangible futures: Combining scenario thinking and personas—A pilot study on urban mobility. *Futures*, Vol. 117/102513, pp. 1–26. https://doi.org/10.1016/j.futures.2020.102513.

Vallet, F., Hörl, S., and Gall, T. (2022) Matching synthetic populations with personas: A test application for urban mobility. DESIGN2022, May 2022, Cavtat, Croatia, pp. 1795–1804. https://doi.org/10.1017/pds.2022.182.

Vandael, S., Claessens, B., Ernst, D., Holvoet, T., and Deconinck, G. (2015) Reinforcement learning of heuristic EV fleet charging in a day-ahead electricity market. *IEEE Transactions on Smart Grid*, Vol. 6/4, pp. 1795–1805. https://doi.org/10.1109/TSG.2015.2393059.

Verloo, N. (2019) Captured by bureaucracy: Street-level professionals mediating past, present and future knowledge, In: Raco, M. and Savini, F. (eds), *Planning and knowledge: How new forms of technocracy are shaping contemporary cities*. Policy Press, pp. 75–89.

Vosooghi, R., Puchinger, J., Jankovic, M., and Vouillon, A. (2019) Shared autonomous vehicle simulation and service design. *Transportation Research. Part C, Emerging Technologies*, Vol. 107, pp. 15–33. https://doi.org/10.1016/j.trc.2019.08.006.

Vosooghi, R., Puchinger, J., Bischoff, J., Jankovic, M., and Vouillon, M. (2020) Shared autonomous electric Vehicle service performance: Assessing the impact of charging infrastructure and battery capacity. *Transportation Research Part D: Transport and Environment*, Vol. 81. https://doi.org/10.1016/j.trd.2020.102283.

Wan, Z., Li, J., He, H., and Prokhorov, D. (2018) Model-free real-time EV charging scheduling based on deep reinforcement learning. *IEEE Transactions on Smart Grid*, Vol. 10/5, pp. 5246–5257. https://doi.org/10.1109/TSG.2018.2879572.

Wang, Z., and Li, F. (2011) Critical peak pricing tariff design for mass consumers in Great Britain. 2011 IEEE Power and Energy Society General Meeting.

Wang, R., Wang, P., Xiao, G., and Gong, S. (2014) Power demand and supply management in microgrids with uncertainties of renewable energies. *International Journal of Electrical Power & Energy Systems*, Vol. 63. pp. 260–269. https://doi.org/10.1016/j.ijepes.2014.05.067.

Wang, D., Coignard, J., Zeng, T., Zhang, C., and Saxena, S. (2016) Quantifying electric vehicle battery degradation from driving vs. vehicle-to-grid services. *Journal of Power Sources*, Vol. 332, pp. 193–203.

Wang, R., Xiao, G., and Wang, P. (2017) Hybrid centralized-decentralized (HCD) charging control of electric vehicles. *IEEE Transactions on Vehicular Technology*, Vol. 66/8, pp. 6728–6741.

Wang, F., Gao, J., Li, M., and Zhao, L. (2020) Autonomous PEV charging scheduling using Dyna-Q reinforcement learning. *IEEE Transactions on Vehicular Technology*, Vol. 69/11, pp. 12609–12620. https://doi.org/10.1109/TVT.2020.3026004.

Watson, V. (2002) Do we learn from planning practice? The contribution of the “practice movement” to planning theory. *Journal of Planning Education and Research*, Vol. 22/2, pp. 178–187. https://doi.org/10.1177/2F0739456X02238446.

Watson, V. (2003) Conflicting rationalities: Implications for planning theory and ethics. *Planning Theory & Practice*, Vol. 4/4, pp. 395–407. https://doi.org/10.1080/1464935032000146318.

Watson, M. (2012) How theories of practice can inform transition to a decarbonised transport system. *Journal of Transport Geography*, Vol. 24, pp. 488–496. https://doi.org/10.1016/j.jtrangeo.2012.04.002.

Wildfire, C. (2018) How can we spearhead city-scale digital twins? *Infrastructure Intelligence.* http://www.infrastructure-intelligence.com/article/may-2018/how-can-we-spearhead-city-scale-digital-twins [accessed August 2023].

Woodcock, J., Edwards, P., Tonne, C., Armstrong, B.G., Ashiru, O., Banister, D., Beevers, S., Chalabi, Z., Chowdhury, Z., Cohen, A., Franco, O.H., Haines, A., Hickman, R., Lindsay, G., Mittal, I., Mohan, D., Tiwari, G., Woodward, A., and Roberts, I. (2009) Public health benefits of strategies to reduce greenhouse-gas emissions: Urban land transport. *The Lancet*, Vol. 374/9705, pp. 1930–1943. https://doi.org/10.1016/S0140-6736(09)61714-1.

Xie, X.-F., Smith, S., and Barlow, G. (2012) Schedule-driven coordination for real-time traffic network control. Proceedings of the International Conference on Automated Planning and Scheduling.

Yameogo, B.F., Vandanjon, P.-O., Gastineau, P., and Hankach, P. (2021) Generating a two-layered synthetic population for French municipalities: Results and evaluation of four synthetic reconstruction methods. *Journal of Artificial Societies and Social Simulation*, Vol. 24/5. https://doi.org/10.18564/jasss.4482.

Yang, Y., Jia, Q.-S., Deconinck, G., Guan, X., Qiu, Z., and Hu, Z. (2019) Distributed Coordination of EV Charging with Renewable Energy in a Microgrid of Buildings. IEEE Power & Energy Society General Meeting (PESGM), Atlanta, pp. 1–1. https://doi.org/10.1109/PESGM40551.2019.8973991.

Yin, Y., Zhou, M., and Li, G. (2015) Dynamic decision model of critical peak pricing considering electric vehicles' charging load. International Conference on Renewable Power Generation (RPG 2015), Beijing, pp. 1–6. https://doi.org/10.1049/cp.2015.0564.

Yüksel, I. (2012) Developing a multi-criteria decision-making model for PESTEL analysis. *International Journal of Business and Management*, Vol. 7/24. https://doi.org/10.5539/ijbm.v7n24p52.

Zhang, X., Liang, Y., and Liu, W. (2017) Pricing model for the charging of electric vehicles based on system dynamics in Beijing. *Energy*, Vol. 119, pp. 218–234. https://doi.org/10.1016/j.energy.2016.12.057.

Ziemke, D., Charlton, B., Hörl, S., and Nagel, K. (2021) An efficient approach to create agent-based transport simulation scenarios based on ubiquitous Big Data and a new, aspatial activity-scheduling model. *Transportation Research Procedia*, Vol. 52, pp. 613–620. https://doi.org/10.1016/j.trpro.2021.01.073.

Printed in the United States
by Baker & Taylor Publisher Services